Sequences and Mathematical Induction

In Mathematical Olympiad and Competitions

Second Edition

Mathematical Olympiad Series

ISSN: 1793-8570

Published

Vol. 16 *Sequences and Mathematical Induction: In Mathematical Olympiad and Competitions Second Edition*
by Zhigang Feng (Shanghai Senior High School, China)
translated by: Feng Ma, Youren Wang

Vol. 15 *Mathematical Olympiad in China (2011–2014): Problems and Solutions*
edited by Bin Xiong (East China Normal University, China) &
Peng Yee Lee (Nanyang Technological University, Singapore)

Vol. 14 *Probability and Expectation*
by Zun Shan (Nanjing Normal University, China)
translated by: Shanping Wang (East China Normal University, China)

Vol. 13 *Combinatorial Extremization*
by Yuefeng Feng (Shenzhen Senior High School, China)

Vol. 12 *Geometric Inequalities*
by Gangsong Leng (Shanghai University, China)
translated by: Yongming Liu (East China Normal University, China)

Vol. 11 *Methods and Techniques for Proving Inequalities*
by Yong Su (Stanford University, USA) &
Bin Xiong (East China Normal University, China)

Vol. 10 *Solving Problems in Geometry: Insights and Strategies for Mathematical Olympiad and Competitions*
by Kim Hoo Hang (Nanyang Technological University, Singapore) &
Haibin Wang (NUS High School of Mathematics and Science, Singapore)

Vol. 9 *Mathematical Olympiad in China (2009–2010)*
edited by Bin Xiong (East China Normal University, China) &
Peng Yee Lee (Nanyang Technological University, Singapore)

The complete list of the published volumes in the series can be found at
http://www.worldscientific.com/series/mos

Vol. 16 | Mathematical Olympiad Series

Sequences and Mathematical Induction

In Mathematical Olympiad and Competitions

Second Edition

Original Author

Feng Zhigang *Shanghai High School, China*

English Translators

Ma Feng *Shanghai High School, China*

Wang Youren *Shanghai High School, China*

Copy Editors

Ni Ming *East China Normal University Press, China*

Kong Lingzhi *East China Normal University Press, China*

Published by

East China Normal University Press
3663 North Zhongshan Road
Shanghai 200062
China

and

World Scientific Publishing Co. Pte. Ltd.
5 Toh Tuck Link, Singapore 596224
USA office: 27 Warren Street, Suite 401-402, Hackensack, NJ 07601
UK office: 57 Shelton Street, Covent Garden, London WC2H 9HE

British Library Cataloguing-in-Publication Data
A catalogue record for this book is available from the British Library.

Mathematical Olympiad Series — Vol. 16
SEQUENCES AND MATHEMATICAL INDUCTION
In Mathematical Olympiad and Competitions
Second Edition

ISBN 978-981-121-103-4
ISBN 978-981-121-207-9 (pbk)

For any available supplementary material, please visit
https://www.worldscientific.com/worldscibooks/10.1142/11572#t=suppl

Printed in Singapore

Introduction

Mathematical induction is an important method used to prove particular math statements and is widely applicable in different branches of mathematics, among which it is most frequently used in sequences. This book is rewritten on the basis of the book *Methods and Techniques for Proving by Mathematical Induction*, and is written with an understanding that sequences and mathematical induction overlap and share similar ideas in the realm of mathematics knowledge. Since there are a lot of theses and books related to this topic already, the author spent quite a lot of time reviewing and refining the contents in order to avoid regurgitating information. For example, this book refers to some of the most updated Math Olympiad problems from different countries, places emphasis on the methods and techniques for dealing with problems, and discusses the connotations and the essence of mathematical induction in different contexts.

The author attempts to use some common characteristics of sequences and mathematical induction to fundamentally connect Math Olympiad problems to particular branches of mathematics. In doing so, the author hopes to reveal the beauty and joy involved with math exploration and at the same time, attempts to arouse readers' interest of learning math and invigorate their courage to challenge themselves with difficult problems.

Preface

Mathematical competitions are a special type of intelligence competition among teenage students. Although many common intelligence competitions are based on science knowledge, mathematical competitions hold the longest history and are the most internationally recognized, and thus, have the biggest impact. China first began to hold math competitions in 1956. The most prestigious and well-known mathematicians from China include Hua Luogeng, Su Buqing, and Jiang Zehan, all of which actively participated in creating and organizing these initial competitions. They were also influential in the publishing of a series of math reading materials for young people and teenagers, which inspired large numbers of young people to begin engaging in mathematical and scientific research. China has participated in the International Math Olympiad since 1986 and received first place awards on a number of occasions. In 1990, China hosted the 31st International Math Olympiad in Beijing, which spoke to China's leading international position and attracted the attention of scientists and educators from many other countries worldwide.

China's success in math competitions over the years has resulted in increasing participation of young people in these competitions across all regions, heightened interest and enthusiasm towards math learning in students, greater facilitation of creative thinking abilities, and improvements in studying habits and efficiency. In addition, they have led to healthy competition strategies to be used in math teaching, which aids in selecting those students with special math talents to participate in the competitions. Those who stand out and achieve success in math competitions prove to have a solid foundation in math, as well as strong science study strategies and skills, and many of the students who are successful in these competitions go on to work in the field of science. In the United States, some winners have gone on to become famous, for example, J. W. Milnor, D. B. Mumford, D. Quillen are all recipients of the Fields Award. In Poland, A. Schinzel, the famous Number Theory expert, received awards in math competitions when he was a student. In Hungary, the famous mathematicians L. Fejér, M. Riesz, G. Szegö, A. Haar, T. Radó were all once winners of math competitions. Hungary was the first country to organize these competitions and as a result, many great mathematicians have come from this region; the number is way beyond the normal ratio of the number of mathematicians to the total population.

Through the implementation of mathematical competitions, participating schools receive a valuable opportunity to strengthen ties between one another and in doing so, exchange math teaching experiences. From this point of view, math competitions become the "catalyst" for math curriculum reforms and become a powerful measure for cultivating excellent talents.

When organizing math competitions, attention should be simultaneously placed on both popularizing the event and improving performance. Popularizing the event is the main focus, as with popularity comes more participation and a lasting, strong influence for the competition, which is the aim of holding these competitions in the first place.

Some may be tempted to become over-concerned with performance instead, organizing and participating in these competitions with a very strong utilitarian objective. These practices are incorrect and are against the original intention of implementing math competitions. These drawbacks have deep social implications and influence and need to be overcome step by step. Math competitions must not be negated because of such drawbacks.

I am very pleased with the publishing of this set of *Mathematical Olympiad Series*. This set of books in particular covers a large range of meticulous topics. Based on my own knowledge and experience, it is rare to come across books of this nature. This set does not only explain the common methods that appear in math competitions, but also provides to-the-point analysis and solutions to the problems, most of which is derived from the authors' own research. This makes this set of books very valuable in preparing for math competitions and can be used as reference materials for students and teachers in primary, middle and high schools.

The authors of this set of books are all teachers and researchers involved in mathematical competitions; many of them are even lead teachers or coaches for the China National Math Olympiad camp and team. They have all contributed to the organization of math competitions in China and in leading China's students to winning achievements and bringing honor to China in IMO. They all put forth many efforts in order to make the publishing of this set of books possible. The East China Normal University Publishing House designed this set of books using their experience in publishing math competition books, such as *Math Olympiad Courses* and *Going Toward IMO*; it is quite evident that they spent a lot of time and energy on it. I am very grateful for the work that the authors and editors put in for this set of books and I would like to conclude by offering my sincerest wishes for a successful future for China in mathematical competitions.

Wang Yuan

Famous Mathematician. Member of the Chinese Academy of Sciences. Former Chair of the Chinese Mathematical Society and Chinese Mathematical Olympiad Committee.

Acknowledgment

Raymond Luo from University of Michigan and Michael Li from MIT helped proofread this book. Mr. Ni and Mr. Kong from ECNU Press supported us during the whole publishing process. Our families, friends and colleagues gave us tremendous support in this project. Taking this opportunity, we would like to give our sincerest thanks to all of them.

Due to our limited understanding on the original work which is in Chinese, plus the given limited time which was mostly the weekends and summer and winter breaks, we probably did not manage to make this translation perfect. We will be very grateful if our readers do not hesitate to point out the mistakes in our translations.

Ma Feng & Wang Youren

August 8, 2019

Notations

$\mathbf{N}$	The set comprised of natural numbers $0, 1, 2, \cdots$
$\mathbf{N}^*$	The set comprised of positive integers $1, 2, \cdots$
$\mathbf{Z}$	The set of integers
$\mathbf{Q}$	The set of rational numbers
$\mathbf{R}$	The set of real numbers
$\mathbf{C}$	The set of complex numbers
$a \mid b$	The integer b is divisible by the integer a
$a \nmid b$	The integer b is not divisible by the integer a
max	The maximum value
min	The minimum value
$[x]$, $\lfloor x \rfloor$	The greatest integer no bigger than the real number x, i.e., the integer part of x
$\lceil x \rceil$	The smallest integer no less than the real number x
$\{x\}$	The decimal part of the real number x, i.e., $\{x\} = x - [x]$
$\sum$	To find the sum
$\prod$	To find the product
$\equiv$	Congruent

Table of Contents

Chapter 1 Knowledge and Technique

1 The First Form of Mathematical Induction

Mathematical induction is a common proof technique used to prove a given proposition $P(n)$ involving a positive integer n. It is a direct corollary of the following axiom of induction.

Axiom of induction Let S be a subset of the set of positive integers $\mathbf{N}^*$, satisfying:

(1) $1 \in S$;

(2) If $n \in S$, then $n+1 \in S$.

Then $S = \mathbf{N}^*$.

Axiom of induction is one of the five axioms for positive integers presented by Peano. The axiom laid the foundation for mathematical induction.

The first form of mathematical induction is the most common form, which is referred to in our high school textbooks.

The first form of mathematical induction Let $P(n)$ be a proposition (or property) about (of) positive integer n. Suppose the following conditions hold.

(1) $P(n)$ is true when $n = 1$;

(2) It can be inferred from the validity of $P(n)$ that $P(n+1)$ is true.

Then $P(n)$ is true for all $n \in \mathbf{N}^*$.

Proof. Let $S = \{n \mid n \in \mathbf{N}^* \text{ and } P(n) \text{ is true.}\}$. Then S is a subset of $\mathbf{N}^*$. Noting (1), we have $1 \in S$; noting (2), if $n \in S$, then $n+1 \in S$. Thus, by axiom of induction, we can deduce that $S = \mathbf{N}^*$, i.e., $P(n)$ is true for all $n \in \mathbf{N}^*$.

Explanation. In fact, the first form of mathematical induction is equivalent to the axiom of induction. So, they are also named as the principle of mathematical induction. The first form of mathematical induction is called mathematical induction for short.

It is not hard for high school students to understand the implications and validity of mathematical induction. However, utilizing mathematical induction is no easy job.

Utilizing mathematical induction is composed of two steps. Checking the validity of $P(1)$ lays the foundation. Combining the inductive hypothesis with relevant knowledge, we gain the recursion of $P(n+1)$. These two steps complement each other in proving the proposition and constitute the logical structure of the inductive proof. Most importantly, it is necessary to make use of the inductive hypothesis in inductive proof, which provides a criterion for the validity of the proof.

Example 1. For any $n \in \mathbf{N}^*$, prove that

$$\frac{1}{1\times 2}+\frac{1}{2\times 3}+\cdots+\frac{1}{n(n+1)}=1-\frac{1}{n+1}. \quad ①$$

Proof. When $n=1$, the left side of ① $=\frac{1}{2}$, while the right side of ① $=1-\frac{1}{1+1}=\frac{1}{2}$. Thus, ① holds for $n=1$.

Now suppose that ① holds for n. Let's consider the statement with $n+1$.

By $\frac{1}{k(k+1)}=\frac{1}{k}-\frac{1}{k+1}$, we have

$$\begin{aligned}&\frac{1}{1\times 2}+\frac{1}{2\times 3}+\cdots+\frac{1}{(n+1)(n+2)}\\&=\left(1-\frac{1}{2}\right)+\left(\frac{1}{2}-\frac{1}{3}\right)+\cdots+\left(\frac{1}{n+1}-\frac{1}{n+2}\right)\\&=1-\frac{1}{n+2}.\end{aligned} \quad ②$$

Therefore ① holds for $n+1$.

In conclusion, by the principle of mathematical induction, we prove that ① holds for all positive integer n.

Explanation. This proof is wrong in that the inductive hypothesis is not made use of when we prove ① holds for $n+1$.

Here is the correct process:

Noting the inductive hypothesis, we have

$$\begin{aligned}&\frac{1}{1\times 2}+\frac{1}{2\times 3}+\cdots+\frac{1}{n(n+1)}+\frac{1}{(n+1)(n+2)}\\ =&\left(1-\frac{1}{n+1}\right)+\frac{1}{(n+1)(n+2)}\\ =&\left(1-\frac{1}{n+1}\right)+\left(\frac{1}{n+1}-\frac{1}{n+2}\right)\\ =&1-\frac{1}{n+2}.\end{aligned}$$

Therefore ① holds for $n+1$.

Actually ② is derived accurately. However, it is a direct proof for ① without the technique of mathematical induction. This mistake is often made by high school students, which must be corrected seriously. Otherwise, it will be hard for students to establish an accurate thinking structure of deduction.

Example 2. Let $n\in\mathbf{N}^*$. Prove that after removing any square from a $2^n\times 2^n$ grid, the remaining part can be tiled with L-shaped "⊥" tiles (no gaps and no overlaps).

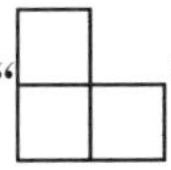

Proof. When $n=1$, since a "⊞" turns into a "⊥" when any square is removed from it. So the proposition holds for $n=1$.

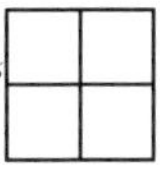

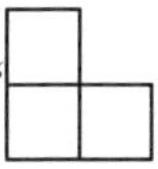

Now we assume that the proposition is true when $n=k$. That is, after removing any square from a $2^k\times 2^k$ grid, the remaining part can be tiled with "⊥" tiles. Let's consider the statement with $n=k+1$.

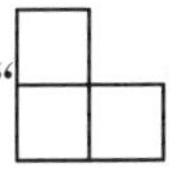

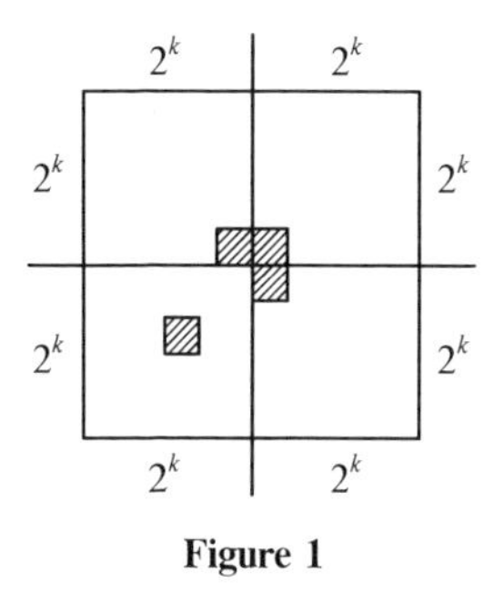

Figure 1

As shown in Figure 1, when we divide the $2^{k+1} \times 2^{k+1}$ grid into four $2^k \times 2^k$ grids along the middle latitude and longitude, the removed square must fall in one of the $2^k \times 2^k$ grids. So, at first, we can put a “ ” tile in the middle of the remaining part. Then let’s remove the four shaded squares, as shown in Figure 1. Now one square is removed in each $2^k \times 2^k$ grid. Noting the inductive hypothesis, all of them can be tiled with “ ” tiles. Along with the tile put in the middle, we get the validity of the proposition about $n = k + 1$.

Generalizing all above, the proposition is true for all positive integers n.

Explanation. This example shows the common expression of the proof technique of mathematical induction. The structure can certainly vary according to one’s own style. However, it is necessary to achieve a correct transition between the inductive hypothesis and conclusion, which is the key step in proving a proposition with mathematical induction.

Example 3. Let x, y be real numbers satisfying that $x + y$, $x^2 + y^2$, $x^3 + y^3$ and $x^4 + y^4$ are all integers. Prove that $x^n + y^n$ is an integer for any $n \in \mathbf{N}^*$.

Proof. This problem calls for a varied form of the mathematical induction: let $P(n)$ be a proposition (or property) about (of) positive integer n. If

(1) $P(n)$ is true when $n = 1$ and 2;

(2) If can be inferred from the validity of $P(n)$ and $P(n+1)$ that $P(n+2)$ is true.

Then $P(n)$ is true for all $n \in \mathbf{N}^*$.

Actually, this change only adjusted the step size in the process of induction. This kind of cases will occur frequently in later discussions.

Let's return to the problem. Since $x+y$ and x^2+y^2 are both integers, the proposition is true when $n=1$ and 2.

Suppose that the proposition is true for n and $n+1$. That is, x^n+y^n and $x^{n+1}+y^{n+1}$ are both integers. Let's consider the statement with $n+2$. Now

$$x^{n+2}+y^{n+2}=(x+y)(x^{n+1}+y^{n+1})-xy(x^n+y^n).$$

So, in order to prove $x^{n+2}+y^{n+2}\in\mathbf{Z}$, with the help of the inductive hypothesis and the condition that $x+y\in\mathbf{Z}$, we need only prove that $xy\in\mathbf{Z}$.

Noting that $x+y$, $x^2+y^2\in\mathbf{Z}$, we have

$$2xy=(x+y)^2-(x^2+y^2)\in\mathbf{Z}.$$

If $xy\notin\mathbf{Z}$, let $xy=\frac{m}{2}$, where m is odd. Since x^2+y^2, $x^4+y^4\in\mathbf{Z}$, we can infer that

$$2x^2y^2=(x^2+y^2)^2-(x^4+y^4)\in\mathbf{Z}.$$

Therefore $2\times\left(\frac{m}{2}\right)^2=\frac{m^2}{2}\in\mathbf{Z}$. However, m is odd. There is a contradiction. So $xy\in\mathbf{Z}$ and, thus, the proposition is true for $n+2$.

In conclusion, $x^n+y^n\in\mathbf{Z}$ for all $n\in\mathbf{N}^*$.

Example 4. Let $\theta\in\left(0,\frac{\pi}{2}\right)$ and n be a positive integer greater than 1. Prove that

$$\left(\frac{1}{\sin^n\theta}-1\right)\left(\frac{1}{\cos^n\theta}-1\right)\geqslant 2^n-2^{\frac{n}{2}+1}+1. \qquad ①$$

Proof. When $n=2$, the left and right sides of ① are equal. So the proposition is true when $n=2$.

Assume that the proposition is true for $n(\geqslant 2)$. Then

$$\left(\frac{1}{\sin^{n+1}\theta}-1\right)\left(\frac{1}{\cos^{n+1}\theta}-1\right)$$
$$=\frac{1}{\sin^{n+1}\theta\cos^{n+1}\theta}(1-\sin^{n+1}\theta)(1-\cos^{n+1}\theta)$$
$$=\frac{1}{\sin^{n+1}\theta\cos^{n+1}\theta}(1-\sin^{n+1}\theta-\cos^{n+1}\theta)+1$$
$$=\frac{1}{\sin\theta\cos\theta}\left(\frac{1}{\sin^{n}\theta\cos^{n}\theta}-\frac{\cos\theta}{\sin^{n}\theta}-\frac{\sin\theta}{\cos^{n}\theta}\right)+1$$
$$=\frac{1}{\sin\theta\cos\theta}\left[\left(\frac{1}{\sin^{n}\theta}-1\right)\left(\frac{1}{\cos^{n}\theta}-1\right)+\frac{1-\cos\theta}{\sin^{n}\theta}+\frac{1-\sin\theta}{\cos^{n}\theta}-1\right]+1$$
$$\geqslant\frac{1}{\sin\theta\cos\theta}\left[(2^{n}-2^{\frac{n}{2}+1})+2\sqrt{\frac{(1-\cos\theta)(1-\sin\theta)}{\sin^{n}\theta\cos^{n}\theta}}\right]+1, \qquad ②$$

where ② is deduced from the inductive hypothesis and the AM-GM inequality.

Note that $\sin\theta\cos\theta=\frac{1}{2}\sin 2\theta\leqslant\frac{1}{2}$, and that

$$\frac{(1-\cos\theta)(1-\sin\theta)}{\sin^{n}\theta\cos^{n}\theta}=\left(\frac{1}{\sin\theta\cos\theta}\right)^{n-2}\cdot\frac{1}{(1+\sin\theta)(1+\cos\theta)},$$

in which

$$\begin{aligned}(1+\sin\theta)(1+\cos\theta)&=1+\sin\theta+\cos\theta+\sin\theta\cos\theta\\&=1+t+\frac{t^{2}-1}{2}\\&=\frac{1}{2}(t+1)^{2}\leqslant\frac{1}{2}(\sqrt{2}+1)^{2}\end{aligned}$$

(we have made use of the property that $t=\sin\theta+\cos\theta=\sqrt{2}\sin\left(\theta+\frac{\pi}{4}\right)\in(1,\sqrt{2}]$).

Hence

$$\sqrt{\frac{(1-\cos\theta)(1-\sin\theta)}{\sin^{n}\theta\cos^{n}\theta}}\geqslant\frac{2^{\frac{n-1}{2}}}{\sqrt{2}+1}=2^{\frac{n}{2}}-2^{\frac{n-1}{2}}.$$

Therefore it can be deduced from ② that

$$\left(\frac{1}{\sin^{n+1}\theta}-1\right)\left(\frac{1}{\cos^{n+1}\theta}-1\right)$$
$$\geqslant 2[(2^n-2^{\frac{n}{2}+1})+2(2^{\frac{n}{2}}-2^{\frac{n-1}{2}})]+1$$
$$=2(2^n-2^{\frac{n+1}{2}})+1=2^{n+1}-2^{\frac{n+1}{2}+1}+1.$$

So the proposition is true for $n+1$.

In conclusion, the proposition is true for all $n\in\mathbf{N}^*$ $(n\geqslant 2)$.

Explanation. The examples above refer to knowledge in several branches ranging from algebra, number theory to combination and demonstrate the diverse applications of mathematical induction.

Example 5. Sequence $\{a_n\}$ is defined as follows:

$$a_1=1,\ a_n=a_{n-1}+a_{[\frac{n}{2}]},\ n=2,3,\cdots.$$

Prove that there are infinitely many terms in the sequence that are multiples of 7.

Proof. Calculating directly by the recurrence formula, we can get that

$$a_1=1,\ a_2=2,\ a_3=3,\ a_4=5,\ a_5=7.$$

Now we suppose that $a_n (n\geqslant 5)$ is a multiple of 7. Let's find a subscript $m>n$ satisfying $7\mid a_m$.

Since $a_n\equiv 0 \pmod 7$, we have $a_{2n}=a_{2n-1}+a_n\equiv a_{2n-1} \pmod 7$ and $a_{2n+1}=a_{2n}+a_n\equiv a_{2n} \pmod 7$. So $a_{2n-1}\equiv a_{2n}\equiv a_{2n+1} \pmod 7$. Let r be the remainder when a_{2n-1} is divided by 7. If $r=0$, it suffices to take $m=2n-1$; if $r\neq 0$, let's consider the following 7 numbers:

$$a_{4n-3},\ a_{4n-2},\ \cdots,\ a_{4n+3}. \quad ①$$

Note that

$$a_{4n-2}=a_{4n-3}+a_{2n-1}\equiv a_{4n-3}+r \pmod 7,$$
$$a_{4n-1}=a_{4n-2}+a_{2n-1}\equiv a_{4n-2}+r \pmod 7\equiv a_{4n-3}+2r \pmod 7,$$
$$a_{4n}=a_{4n-1}+a_{2n}\equiv a_{4n-1}+r\equiv a_{4n-3}+3r,\ \cdots,$$
$$a_{4n+3}=a_{4n+2}+a_{2n+1}\equiv a_{4n+2}+r\equiv a_{4n-3}+6r \pmod 7.$$

Therefore a_{4n-3}, a_{4n-2}, $\cdots$, a_{4n+3} constitutes a complete system of residues of modulo 7. So, there exists an $m \in \{4n-3, 4n-2, \cdots, 4n+3\}$ satisfying that $a_m \equiv 0 \pmod 7$.

In this way, starting from a_5 and combining with the deduction above, we proved that there exist infinitely many terms in the sequence that are multiples of 7.

Example 6. (1) For any positive integer $n (\geqslant 2)$, prove that there exist n different positive integers a_1, $\cdots$, a_n, satisfying

$$(a_i - a_j) \mid (a_i + a_j),$$

for any $1 \leqslant i < j \leqslant n$.

(2) Is there an infinite set $\{a_1, a_2, \cdots\}$ of positive integers satisfying $(a_i - a_j) \mid (a_i + a_j)$ for any $i \neq j$?

Proof. (1) When $n = 2$, it suffices to take 1 and 2.

Suppose that the proposition is true for n. That is, there exist positive integers $a_1 < a_2 < \cdots < a_n$, satisfying $(a_i - a_j) \mid (a_i + a_j)$, for any $1 \leqslant i < j \leqslant n$. Now we consider the following $n+1$ numbers:

$$A, A + a_1, A + a_2, \cdots, A + a_n. \qquad ①$$

Where $A = a_n!$ and $a_n! = 1 \times 2 \times 3 \times \cdots \times a_n$.

Take two numbers $x < y$ from ①. If $x = A$, $y = A + a_i$, $1 \leqslant i \leqslant n$, then $y - x = a_i$ and $x + y = 2A + a_i$. Combining with $a_i \leqslant a_n$, we have $a_i \mid A$. So $(y - x) \mid (y + x)$; if $x = A + a_i$, $y = A + a_j$, $1 \leqslant i < j \leqslant n$, then $y - x = a_j - a_i$, $y + x = 2A + (a_i + a_j)$. By inductive hypothesis that $(a_j - a_i) \mid (a_j + a_i)$, noting $a_j - a_i < a_n$, we have $(a_j - a_i) \mid A$. Therefore $(y - x) \mid (y + x)$. Thus, the proposition is true for $n + 1$.

In conclusion, for any $n \in \mathbf{N}^*$, $n \geqslant 2$, there exist n positive integers satisfying the conditions.

(2) If there exist infinitely many positive integers $a_1 < a_2 < \cdots$, satisfying $(a_i - a_j) \mid (a_i + a_j)$ for any $1 \leqslant i < j$, then for any $j > 1$, we have $(a_j - a_1) \mid (a_j + a_1)$. So $(a_j - a_1) \mid 2a_1$. However, since $a_1 <$

$a_2 < \cdots$, we infer that $2a_1$ is divisible by infinitely many positive integers, which is a contradiction. Thus, there can't be infinitely many positive integers satisfying the conditions.

Explanation. What mathematical induction proves is that for any $n \in \mathbf{N}^*$, $P(n)$ is true. That is to say, it deals with propositions about any limited positive integer n rather than $P(\infty)$. Here we demonstrate partly the essential difference between finiteness and infinity by the comparison of (1) and (2) in the example.

We can certainly deal with some results concerning infinity by mathematical induction, such as what we have done to Example 5. Comparing the structure of the recursion in Example 5 with the one in Example 6, we can find the essential difference between them. The former is compatible with the previous result, while the latter isn't.

2 The Second Form of Mathematical Induction

The second form of mathematical induction Let $P(n)$ be a proposition (or property) about (of) positive integer n. Suppose the following conditions hold.

(1) $P(n)$ is true when $n = 1$;

(2) If $P(k)$ is true for all positive integers k less than n, we can infer that $P(n)$ is true.

Then $P(n)$ is true for all $n \in \mathbf{N}^*$.

Proof. Consider proposition $Q(n)$: " for all $1 \leqslant k \leqslant n$, $k \in \mathbf{N}^*$, $P(k)$ is true." It can be deduced from the validity of $Q(n)$ that $P(n)$ is true.

When $n = 1$, by (1), we have that $Q(n)$ is true.

Now we suppose that $Q(n-1)$ $(n \geqslant 2)$ holds. That is, $P(k)$ is true for all $1 \leqslant k \leqslant n-1$. Then by (2), we have $P(n)$ is true. Thus, for all $1 \leqslant k \leqslant n$, $k \in \mathbf{N}^*$, $P(k)$ is true and therefore $Q(n)$ is true.

Hence by the first form of mathematical induction, we can deduce that for all $n \in \mathbf{N}^*$, $Q(n)$ is true and furthermore, $P(n)$ is true. Thus, we have proved the validity of the second form of mathematical induction.

The second form of mathematical induction is a corollary of the first form. When putting forward the inductive hypothesis, we suppose that $P(1), \cdots, P(n-1)$ are true. Then we prove that $P(n)$ is true under these conditions. This is where the second form differs from the first form, and this difference may sometimes be very handy in proving a proposition.

Example 1. The sequence of real numbers $a_1, a_2, \cdots$ satisfying that $a_{i+j} \leqslant a_i + a_j$ for all $i, j \in \mathbf{N}^*$. For any $n \in \mathbf{N}^*$, prove that

$$a_1 + \frac{a_2}{2} + \frac{a_3}{3} + \cdots + \frac{a_n}{n} \geqslant a_n. \qquad ①$$

Proof. When $n = 1$, it is obvious that the proposition is true.

Now we suppose ① holds for all positive integers less than n. That is, for $1 \leqslant k \leqslant n-1$, we have

$$a_1 + \frac{a_2}{2} + \cdots + \frac{a_k}{k} \geqslant a_k.$$

Let $b_k = a_1 + \frac{a_2}{2} + \cdots + \frac{a_k}{k}$, $k = 1, 2, \cdots, n-1$. By the hypothesis, we have $\sum_{k=1}^{n-1} b_k \geqslant \sum_{k=1}^{n-1} a_k$, equivalently,

$$(n-1)a_1 + \frac{n-2}{2}a_2 + \cdots + \frac{n-(n-1)}{n-1}a_{n-1} \geqslant \sum_{k=1}^{n-1} a_k.$$

Adding $\sum_{k=1}^{n-1} a_k$ to both sides, we have

$$n\left(a_1 + \frac{a_2}{2} + \cdots + \frac{a_{n-1}}{n-1}\right) \geqslant 2\sum_{k=1}^{n-1} a_k.$$

Hence

$$n\left(a_1 + \frac{a_2}{2} + \cdots + \frac{a_n}{n}\right) \geqslant a_n + 2\sum_{k=1}^{n-1} a_k. \qquad ②$$

It can be deduced from the condition that

$$a_1 + a_{n-1} \geqslant a_n,\ a_2 + a_{n-2} \geqslant a_n,\ \cdots,\ a_{n-1} + a_1 \geqslant a_n.$$

Thus, $2\sum_{k=1}^{n-1} a_k \geqslant (n-1)a_n$. Then by ②, we deduce that ① holds for n.

Therefore for any $n \in \mathbf{N}^*$, the inequality ① holds.

Example 2. The sequence of positive integers $c_1, c_2, \cdots$ satisfies the following condition. For any positive integers m, n, if $1 \leqslant m \leqslant \sum_{i=1}^{n} c_i$, there exist positive integers $a_1, a_2, \cdots, a_n$ satisfying that

$$m = \sum_{i=1}^{n} \frac{c_i}{a_i}.$$

Then for any given $i \in \mathbf{N}^*$, what is the maximal c_i?

Proof. Let's prove that the maximal c_1 is 2 and the maximal c_i is $4 \times 3^{i-2}$, when $i \geqslant 2$.

For this purpose, we need first prove that $c_1 \leqslant 2$ and $c_i \leqslant 4 \times 3^{i-2}$, when $i \geqslant 2$.

In fact, if $c_1 > 1$, letting $(m, n) = (c_1 - 1, 1)$, we have that there exists an $a_1 \in \mathbf{N}^*$ satisfying that $c_1 - 1 = \frac{c_1}{a_1}$. Then

$$a_1 = \frac{c_1}{c_1 - 1} = 1 + \frac{1}{c_1 - 1}.$$

a_1 is an integer only when $c_1 = 2$. So $c_1 \leqslant 2$.

Now we suppose that ① holds for all $i = 1, 2, \cdots, k-1 (k \geqslant 2)$. Let $(m, n) = (c_k, k)$. Then there exist $a_1, \cdots, a_k \in \mathbf{N}^*$ satisfying $c_k = \frac{c_1}{a_1} + \cdots + \frac{c_k}{a_k}$. This calls for $a_k \geqslant 2$, otherwise $\sum_{i=1}^{k-1} \frac{c_i}{a_i} = 0$ which is contrary to that a_i and c_i are positive integers. Thus, $c_k \leqslant \frac{c_k}{2} + \sum_{i=1}^{k-1} c_i$, equivalently, $c_k \leqslant 2\sum_{i=1}^{k-1} c_i$. Hence

$$c_k \leqslant 2(2 + 4 + 4 \times 3 + \cdots + 4 \times 3^{k-3}) = 4 \times 3^{k-2}.$$

Therefore by the second form of mathematical induction, we can deduce that ① holds.

When $c_1 = 2$, $c_i = 4 \times 3^{i-2} (i \geqslant 2)$, let's prove that the sequence $\{c_i\}$ has the property in the problem. ②

Let's induct on n. When $n = 1$, $m \leqslant c_1 = 2$. So $m = 1$ or 2. If $m = 1$, it suffices to let $a_1 = 2$. If $m = 2$, it suffices to let $a_1 = 1$.

Now we suppose the sequence $\{c_i\}$ has the property given in the problem for 1, 2, ⋯, $n - 1$. Let's consider the situation for n. Then $1 \leqslant m \leqslant \sum_{i=1}^{n} c_i$.

If $m = 1$, it suffices to let $a_i = nc_i$, $i = 1, 2, \cdots, n$;

If $2 \leqslant m \leqslant \frac{c_n}{2} + 1 = (\sum_{i=1}^{n-1} c_i) + 1$, let $a_n = c_n$ and apply the inductive hypothesis to $m - \frac{c_n}{a_n} = m - 1$. Then we can deduce that ② is true;

If $\frac{1}{2} c_n + 1 < m \leqslant c_n$, it suffices to let $a_n = 2$ and apply the inductive hypothesis to $m - \frac{c_n}{2}$;

If $c_n < m \leqslant \sum_{i=1}^{n} c_i$, it suffices to let $a_n = 1$ and apply the inductive hypothesis to $m - c_n$.

Hence ② is true.

Generalizing all above, when $i \geqslant 2$, the maximal c_i is 2 and the maximal c_i is $4 \times 3^{i-2}$.

Explanation. Comparing the two examples, we can find that there are two ideas when solving problems by the second form of mathematical induction. One is handling the problem as a whole, just as adding the $n - 1$ inequalities in the inductive hypothesis in Example 1. Another is including the situation for n in some situation for 1, 2, ⋯, $n - 1$, which has been demonstrated in the latter part of Example 2.

Example 3. $p(x)$ is a real polynomial of degree n and a is a real number no less than 3. Prove that there is at least one number among the following $n+2$ ones that is no less than 1.

$$|a^0-p(0)|,\ |a^1-p(1)|,\ \cdots,\ |a^{n+1}-p(n+1)|$$

Proof. Induct on the degree n of $p(x)$.

When $n=0$, $p(x)$ is a constant polynomial. Let $p(x)=c$. From the inequality that $|1-c|+|a-c|\geqslant|a-1|\geqslant 2$, it can be deduced that $\max\{|1-c|,\ |a-c|\}\geqslant 1$. Thus, the proposition is true when $n=0$.

Suppose that the proposition is true for all polynomials with degree less than n. Now we consider the polynomial $p(x)$ of degree n.

Let $f(x)=\frac{1}{a-1}[p(x+1)-p(x)]$. Then the degree of $f(x)$ is less than n. By the inductive hypothesis, we have that there exists $m\in\{0,\ 1,\ 2,\ \cdots,\ n\}$ satisfying that $|a^m-f(m)|\geqslant 1$, i.e.,

$$\left|a^m-\frac{1}{a-1}[p(m+1)-p(m)]\right|\geqslant 1.$$

Therefore

$$|a^{m+1}-p(m+1)+p(m)-a^m|\geqslant a-1\geqslant 2.$$

Hence $\max\{|a^{m+1}-p(m+1)|,\ |a^m-p(m)|\}\geqslant 1$. Then there exists $r\in\{0,\ 1,\ 2,\ \cdots,\ n+1\}$ satisfying that $|a^r-p(r)|\geqslant 1$. Thus, the proposition is true for n.

Generalizing all above, the proposition is true for any polynomial $p(x)$ of degree n.

Explanation. We often make use of the second form of mathematical induction when inducting on the degree of a polynomial. The degree of the difference between two nth degree polynomial may not be $n-1$, however, it must be less than n. We can avoid this kind of discussion by applying the second form of mathematical induction.

Example 4. Prove that any convex n-gon can be overlapped by a triangle spanned by three of its sides or a parallelogram spanned by four of its sides.

Proof. Let's induct on n.

When $n = 3$, the case is trivial. When $n = 4$, if the quadrilateral is a parallelogram, the proposition is true. If it is not a parallelogram, there is a pair of its opposite sides that is not parallel. Extend the two sides. They intersect. Along with one of the rest two sides, they constitute a triangle which overlaps the quadrilateral (see Figure 2).

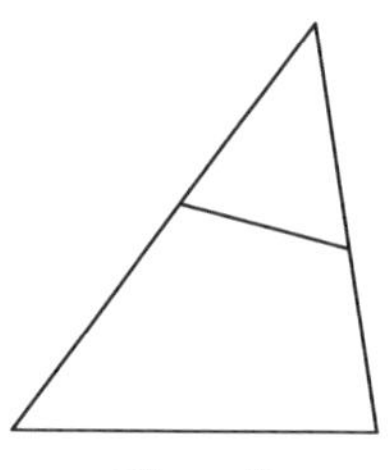

Figure 2

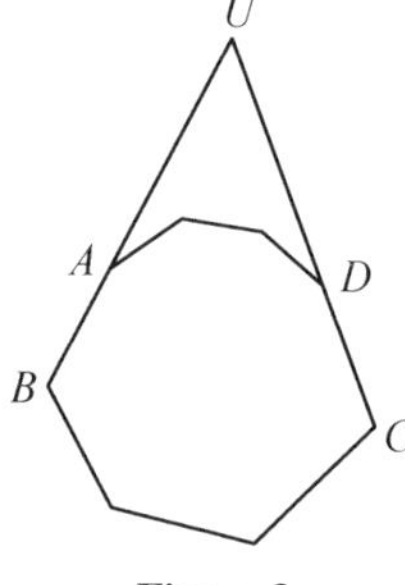

Figure 3

Now we suppose that the proposition is true for any convex m-gon. Here $m < n$ and $n \geqslant 5$. Take an arbitrary side AB of the convex n-gon M. There are at least $n - 3 \geqslant 5 - 3 = 2$ sides except AB and its two adjacent sides. One of the two sides must be unparallel to AB (because there is at most one side parallel to AB). Let the side be CD. Extend BA and CD (without loss of generality, we can suppose the polygon is shown in Figure 3). They intersect at U. Now we substitute the broken line BUC for broken line AD, side BA and side CD which are overlapped by $\angle BUC$. Then we get a convex polygon M_1 which overlaps M. The number of sides of M_1 is less than n. By inductive hypothesis, we can deduce that the proposition is true for n.

Generalizing all above, the proposition is true.

Explanation. Mathematical induction is also widely applied in plane geometry. The proposition in the example, in fact, can be strengthened: if the convex n-gon is not a parallelogram, then it can be overlapped by a triangle spanned by three of its sides.

Example 5. Let $a_1, a_2, \cdots, a_n$ be the first row of an inverted triangle where $a_i \in \{0, 1\}$, $i = 1, 2, \cdots, n$. $b_1, b_2, \cdots, b_{n-1}$ are the second row of the inverted triangle satisfying that if $a_k = a_{k+1}$, then $b_k = 0$; if $a_k \neq a_{k+1}$, then $b_k = 1$, $k = 1, 2, \cdots, n-1$. The remaining

$n-2$ rows of the inverted triangle are similarly defined. Now, what is the maximal number of 1s in the inverted triangle?

Proof. Let f_n be the maximal number of 1s in the inverted triangle. It is easy to see that $f_1 = 1$, $f_2 = 2$, $f_3 = 4$. The examples are

$$1, \quad \begin{matrix} 1 & & 1 \\ & 0 & \end{matrix}, \quad \begin{matrix} 1 & & 1 & & 0 \\ & 0 & & 1 & \\ & & 1 & & \end{matrix}.$$

We can start with the number of 0s in the first line and get the results above. However, it becomes hard to begin from the first line when n grows bigger. When trying to deal with the cases where $n=5$, 6, we can find many 1s in the following table.

$$\begin{matrix} 1 & 1 & 0 & 1 & 1 & 0 & 1 & 1 & 0 & \cdots \\ & 0 & 1 & 1 & 0 & 1 & 1 & 0 & 1 & \cdots \\ & & 1 & 0 & 1 & 1 & 0 & 1 & 1 & \cdots \\ & & & 1 & 1 & 0 & 1 & 1 & 0 & \cdots \\ & & & & & \cdots & & & & \end{matrix}$$

One feature in the table above is that each line recurs every three lines (in a smaller scale). We are thus, simulated to utilize mathematical induction to find the value of f_n.

First, let's prove a lemma.

Lemma. When $n \geqslant 3$, consider the upper three lines of the inverted triangle.

$$\begin{matrix} a_1, a_2, a_3, \cdots, a_n \\ b_1, b_2, \cdots, b_{n-1} \\ c_1, \cdots, c_{n-2} \end{matrix}$$

There are at least $n-1$ 0_s in the three lines.

Proof. Let's induct on n.

The verification of the initial cases is left to readers. Let's see how to complete the process of induction. Note that, in the sense of mod 2, the first three lines are

$$a_1, a_2, a_3, a_4, a_5, \cdots, a_n$$
$$a_1+a_2, a_2+a_3, a_3+a_4, \cdots, a_{n-1}+a_n$$
$$a_1+a_3, a_2+a_4, \cdots, a_{n-2}+a_n$$

If a_1, a_1+a_2, a_1+a_3 are not all 1, then we can remove the three numbers and turn the case into the one of $n-1$. By the inductive hypothesis, we can deduce that the lemma is true.

If $a_1=a_1+a_2=a_1+a_3=1$, then $a_1=1$, $a_2=a_3=0$. Now the beginning part of the first three lines are

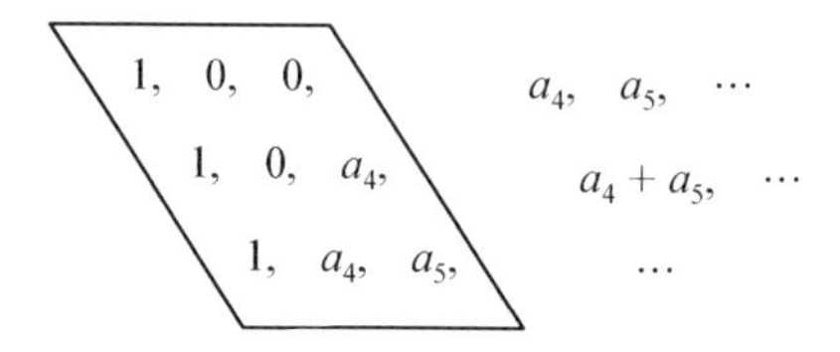

There are at least three 0s among the nine numbers in the parallelogram. Hence removing the nine numbers, by the inductive hypothesis, we can deduce that the lemma is true.

It can be inferred from the lemma that $f_n \leqslant 2(n-1)+f_{n-3}$, $n \geqslant 4$. Since $f_1=1$, $f_2=2$, $f_3=4$, we have $f_n \leqslant \left\lceil \frac{n(n+1)}{3} \right\rceil$, where $\lceil x \rceil$ denote the smallest integer greater than x. Noting the previous example, we can deduce that $f_n = \left\lceil \frac{n(n+1)}{3} \right\rceil$.

Thus, the maximal number of 1s in the inverted triangle is $\left\lceil \frac{n(n+1)}{3} \right\rceil$.

Explanation. Finding an example is a key point in solving this problem. However, it is not hard after making some effort. The difficulties lie in taking every three lines as a whole when dealing with the problem — the spark can come from toying with the example graph above.

Example 6. Let $n \in \mathbf{N}^*$ and function f: $\{1, 2, 3, \cdots, 2^{n-1}\} \to \mathbf{N}^*$ satisfies that $1 \leqslant f(i) \leqslant i$, for $1 \leqslant i \leqslant 2^{n-1}$. Prove that there exists a positive integer sequence $a_1, a_2, \cdots, a_n$, satisfying that $1 \leqslant a_1 < a_2 < \cdots < a_n \leqslant 2^{n-1}$ and $f(a_1) \leqslant \cdots \leqslant f(a_n)$.

Proof. Induct on n.

When $n = 1$, it is obvious that the proposition is true.

Suppose that the proposition is true for $1, 2, \cdots, n-1$. Now we consider the case of n.

For $1 \leqslant i \leqslant 2^{n-1}$, let $t(i)$ denote the maximal m satisfying that there exists a positive integer sequence $i = a_1 < a_2 < \cdots < a_m \leqslant 2^{n-1}$ such that

$$f(a_1) \leqslant f(a_2) \leqslant \cdots \leqslant f(a_m).$$

If the proposition is not true for n, then from the fact that

$$t(1) = \max_{1 \leqslant i \leqslant 2^{n-1}} t(i),$$

it can be deduced that for any $1 \leqslant i \leqslant 2^{n-1}$, we have $t(i) \leqslant n-1$. Let

$$A_j = \{i \mid 1 \leqslant i \leqslant 2^{n-1}, t(i) = j\}, j = 1, 2, \cdots, n-1.$$

Then any two A_j don't intersect and $\bigcup_{j=1}^{n-1} A_j = \{1, 2, 3, \cdots, 2^{n-1}\}$.

Therefore $\sum_{j=1}^{n-1} |A_j| = 2^{n-1}$.

Now for any $1 \leqslant j \leqslant n-1$, let's prove that $|A_j| \leqslant 2^{n-j-1}$.

In fact, if there exists a j satisfying $|A_j| > 2^{n-j-1}$, then there exist $1 \leqslant i_1 < i_2 < \cdots < i_r \leqslant 2^{n-1}$ satisfying that $t(i_1) = t(i_2) = \cdots = t(i_r) = j$, where $r = 2^{n-j-1} + 1$. Now for any $1 \leqslant p < q \leqslant r$, we have $f(i_p) > f(i_q)$ (otherwise, if $f(i_p) \leqslant f(i_q)$, then putting $f(i_p)$ in the front of the increasing sequence f beginning from i_q leads to $t(i_p) \geqslant t(i_q) + 1$, which is contradictory). Hence $f(i_1) > f(i_2) > \cdots > f(i_r)$. Furthermore $f(i_1) \geqslant r = 2^{n-j-1} + 1$. Noting that $1 \leqslant f(i_1) \leqslant i_1$, we have that $i_1 \geqslant 2^{n-j-1} + 1$.

Now from the definition of $t(i_1)$, it can be deduced that there exist $i_1 = a_1 < \cdots < a_j \leqslant 2^{n-1}$, satisfying that $f(a_1) \leqslant \cdots \leqslant f(a_j)$. By the inductive hypothesis, there exist $1 \leqslant b_1 < \cdots < b_{n-j} \leqslant 2^{n-1-j} < i_1 = a_1$ in $\{1, 2, 3, \cdots, 2^{n-j-1}\}$ satisfying that $f(b_1) \leqslant \cdots \leqslant f(b_{n-j})$. Noting that

$$f(b_{n-j}) \leqslant b_{n-j} \leqslant 2^{n-1-j} < r \leqslant f(i_1) = f(a_1),$$

we have

$$1 \leqslant b_1 < \cdots < b_{n-j} < a_1 < a_2 < \cdots < a_j \leqslant 2^{n-1}$$

and

$$f(b_1) \leqslant \cdots \leqslant f(b_{n-j}) \leqslant f(a_1) \leqslant \cdots \leqslant f(a_j),$$

which is contrary to the assumption that the proposition is not true for n. Hence $|A_j| \leqslant 2^{n-j-1}$.

However, we can now infer that

$$2^{n-1} = \sum_{j=1}^{n-1} |A_j| \leqslant \sum_{j=1}^{n-1} 2^{n-1-j} = 1 + 2 + \cdots + 2^{n-2} = 2^{n-1} - 1.$$

This is contradictory. Thus, the proposition is true for n.

Generalizing all above, the proposition is true.

Explanation. Here we utilized proof by contradiction when proving that the proposition is true for n. Other methods of proving can be supplemented when we prove a proposition by mathematical induction.

3 Well-ordering Principle and Infinite Descent

Well-ordering Principle is often applied in mathematical competitions. Its typical form is as follows.

Well-ordering Principle There must be a least element of any nonempty subset T of the set of all positive integers $\mathbf{N}^*$. That is, there exists a positive integer $t_0 \in T$, satisfying that for any $t \in T$, we have $t_0 \leqslant t$.

Proof. Consider set $S = \{x \mid x \in \mathbf{N}^*, x \notin T\}$. It's easy to see that $S = \mathbf{N}^* \setminus T$.

If there is no least integer in T, let's prove that every positive integer belongs to S and thus, $T = \varnothing$, which is contradictory.

First, $1 \in S$. Otherwise, we have that $1 \in T$ and then 1 is the least element of T.

Next, suppose that $1, 2, \cdots, n \in S$, i.e. $1, 2, \cdots, n$ are not elements of T. If $n+1 \in T$, then $n+1$ is the least element of T, which is contradictory. Hence $n+1 \in S$.

Therefore by the second form of mathematical induction, we can deduce that for any $n \in \mathbf{N}^*$, $n \in S$.

Thus, the Well-ordering Principle is true.

When dealing with specific problems, we often make use of some other forms or corollaries of the principle.

1. Greatest Integer Principle. Let M be a nonempty subset of the set of positive integers $\mathbf{N}^*$. Suppose that there is an upper bound of M. That is, there exists $a \in \mathbf{N}^*$ satisfying that for any $x \in M$, $x \leqslant a$. Then M has a greatest element.

2. There is a least element and a greatest element in any finite set of real numbers.

3. The axiom of order. Set M of n real numbers can be written as $M = \{x_1, \cdots, x_n\}$, where $x_1 < x_2 < \cdots < x_n$.

The Well-ordering Principle guides us to start from the extremes (the least or the greatest element) when dealing with a problem. It contains the idea of turning back. We should turn back to the essence of the problem.

Infinite descent comes from solving indeterminate equations. Fermat made use of this method about 400 years ago when proving that there is no positive integer solution to $x^4 + y^4 = z^4$. It's basic idea is as follows:

"If proposition $P(n)$ of positive integer n is true for $n = n_0$, then for some $n_1 \in \mathbf{N}^*$, $n_1 < n_0$, we can prove that the proposition $P(n)$ is true as well." Hence $P(n)$ is not true for any $n \in \mathbf{N}^*$.

This is one form of the Well-ordering Principle, which is often made use of when we deal with problems of number theory, especially indeterminate equations.

Example 1. Given n arbitrary different points on a plane, prove that there exists a circle passing through two of the points that keeps the other $n-2$ points outside.

Proof. Since there are only C_n^2 distances between every two of the n points, there must be two points (say A and B) whose distance is the least (if there are more than one pair of such points, just take any one of them).

Now let's consider the circle P which takes the segment AB as diameter. Then for any point C in P, the longest side of triangle ABC is AB. Noting that AB is least, we can deduce that other $n-2$ points stay outside of circle P. The proposition is proved.

Example 2. Prove that there are no rational numbers x, y, z satisfying that

$$x^2+y^2+z^2+3(x+y+z)+5=0. \qquad ①$$

Proof. Multiply both sides of ① by 4 and complete the squares. We have

$$(2x+3)^2+(2y+3)^2+(2z+3)^2=7.$$

If there exist three rational numbers x, y, z satisfying ①, then there exists integer solution (a, b, c, m) to indeterminate equations

$$a^2+b^2+c^2=7m^2 \qquad ②$$

satisfying that $m>0$.

If there exists integer solution (a_0, b_0, c_0, m_0), $m_0>0$ to ②, we can prove that there is integer solution (a_1, b_1, c_1, m_1), $m_1>0$ to ② and $m_1<m_0$. Thus, by the idea of infinite descent, we can find a decreasing sequence of positive integers $m_0>m_1>m_2>\cdots$, which leads to a contradiction.

In fact, if m_0 is odd, then $m_0^2\equiv 1 \pmod 8$ and thus, $a_0^2+b_0^2+c_0^2\equiv 7 \pmod 8$. However, the square of an integer $\equiv 0, 1, 4 \pmod 8$. Hence $a_0^2+b_0^2+c_0^2\equiv 0, 1, 2, 3, 4, 5, 6 \pmod 8$. $a_0^2+b_0^2+c_0^2\equiv 7 \pmod 8$ can never occur. This leads to contradiction and therefore m_0 is even. Now we have $a_0^2+b_0^2+c_0^2=7m_0^2\equiv 0 \pmod 4$. On the other hand, the square of an integer $\equiv 0$, $1 \pmod 4$. Hence a_0, b_0 and c_0 must be all even. Then let $a_1 = \frac{1}{2}a_0$, $b_1 = \frac{1}{2}b_0$, $c_1 = \frac{1}{2}c_0$, $m_1 = \frac{1}{2}m_0$. We get a solution (a_1, b_1, c_1, m_1) satisfying that $0 < m_1 < m_0$.

In conclusion, there is no rational solution to ①.

Explanation. Starting from the least element or finding a less one from a certain element is an important idea we are introducing in this section. They are fundamentally special forms of mathematical induction. This reflects, to some extent, difficulties and challenges in mastering mathematical induction, probably by which people are attracted to learn math.

Example 3. Let P_1, P_2, $\cdots$, P_n be n noncollinear points on a plane. Prove that there exists at least one line passing through exactly two of the points.

Proof. This is the well known Sylvester Theorem. There are many ways to prove it. One of the brief ways is given with the help of the Well-ordering Principle.

Let's consider the lines P_iP_j passing through at least two of P_1, $\cdots$, P_n. The distances from the points not on the lines to them are greater than 0. There are only finite number of distances (since there are at most C_n^2 lines and finite number of points not on the lines). Hence there is a least value of the distances.

Without loss of generality, let the distance from P_1 to P_2P_3 be the least one. Let's prove that there are no other points among P_1, $\cdots$, P_n on P_2P_3.

If there is another point among P_1, $\cdots$, P_n on line P_2P_3, let P_4 be on P_2P_3. Let Q be the projection of P_1 onto P_2P_3. Then there must be two of P_2, P_3, P_4 on the same side of Q. Let P_2 and P_3 be on the same side of Q. Suppose $|QP_2| < |QP_3|$ (as shown in Figure 4). Then the distance from P_2 to line P_1P_3 is less than or

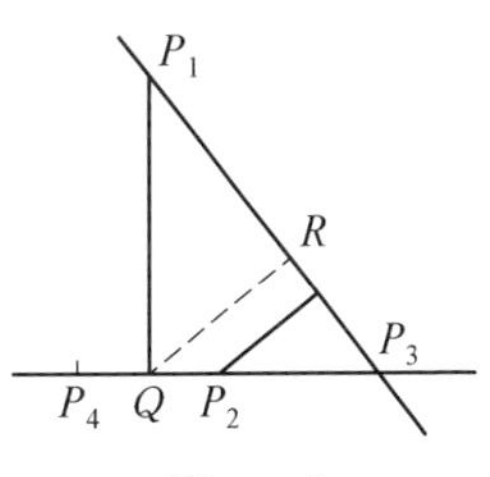

Figure 4

equal to the distance QR from Q to P_1P_3 and $QR < P_1Q$, which is contrary to the minimality of P_1Q. Hence there are no other points on P_2P_3.

Therefore the proposition is true.

Example 4. Prove that there is no positive integer solution to the indeterminate equation

$$x^4 + y^4 = z^4.$$

Proof. Suffice it to prove that there is no positive integer solution to

$$x^4 + y^4 = z^2. \quad ①$$

If there is, suppose that (x, y, z) is a positive integer solution to ①, satisfying that z is the least in all solutions. Now, let d be the greatest common divisor of x and y, i.e., $(x, y) = d$.

Then $d^2 \mid (x^4 + y^4)$ and thus, $d^2 \mid z^2$ and $d \mid z$. Hence $d = 1$ $\left(\text{otherwise, } \left(\frac{x}{d}, \frac{y}{d}, \frac{z}{d}\right) \text{ is also a solution to ①}\right)$. Therefore (x^2, y^2, z) is also a primitive root of

$$u^2 + v^2 = w^2. \quad ②$$

Let y^2 be even. It can be deduced from the general solution of ② that there exist $a, b \in \mathbf{N}^*$, one of which is odd while the other even. a and b satisfy that $(a, b) = 1$ and

$$x^2 = a^2 - b^2, \ y^2 = 2ab, \ z = a^2 + b^2.$$

Noting that y^2 is even, we have that x is odd. Noting that $x^2 + b^2 = a^2$, we can deduce that there exist $m, n \in \mathbf{N}^*$, one of which is odd while the other even. m and n satisfy that $(m, n) = 1$ and

$$x = m^2 - n^2, \ b = 2mn, \ a = m^2 + n^2.$$

Then $y^2 = 4mn(m^2 + n^2)$. Since $(m, n) = 1$,

$$(m, m^2 + n^2) = (n, m^2 + n^2) = 1.$$

Hence $m^2 + n^2$, m, n are perfect squares.

Then let $m = r^2$, $n = s^2$, $m^2 + n^2 = z_1^2$, r, s, $z_1 \in \mathbf{N}^*$. We have $r^4 + s^4 = z_1^2$, where $z_1^2 = a < z$. This is contrary to that z in (x, y, z) is the least in all solutions.

Therefore there is no positive integer solution to ①. The proposition is, thus, proved.

Explanation. We applied another form of infinite descent: "If proposition $P(n)$ is true for some $n \in \mathbf{N}^*$, let n_0 be the least integer satisfying that $P(n)$ is true (the existence of n_0 is given by the Well-ordering Principle). Then we can prove that there exists $n_1 \in \mathbf{N}^*$, $n_1 < n_0$ satisfying that $P(n_1)$ is true." Hence for any $n \in \mathbf{N}^*$, $P(n)$ is not true.

Example 5. Let n be a given positive integer. Is there a finite set of nonzero plane vectors who has more than $2n$ elements and satisfies the following conditions?

(1) For any n vectors in M, we can find n other vectors in M satisfying that the sum of the $2n$ vectors is zero;

(2) For any n vectors in M, we can find $n - 1$ other vectors in M satisfying that the sum of the $2n - 1$ vectors is zero.

Solution. There is no such set M.

In fact, if there exists such set M, since set M is finite, there are finite number of ways to choose n vectors. Therefore there exists a way to choose vectors so that the length of the sum of the n vectors takes maximum. Let the n vectors be $\boldsymbol{u}_1$, $\boldsymbol{u}_2$, $\cdots$, $\boldsymbol{u}_n$ and $\boldsymbol{u}_1 + \boldsymbol{u}_2 + \cdots + \boldsymbol{u}_n = \boldsymbol{s}$.

Draw a line l perpendicular to $\boldsymbol{s}$ from the origin. Then l divide the plane in two. Let M_1 be the set of vectors on the same side of $\boldsymbol{s}$ belonging to M and M_2 be the set of vectors on the other side of $\boldsymbol{s}$ or on l belonging to M. Then $M_1 \cap M_2 = \varnothing$, $M_1 \cup M_2 = M$.

By condition (2), we have that there exist vectors $\boldsymbol{v}_1$, $\cdots$, $\boldsymbol{v}_{n-1}$ in M satisfying that $\boldsymbol{u}_1 + \cdots + \boldsymbol{u}_n + \boldsymbol{v}_1 + \cdots + \boldsymbol{v}_{n-1} = \mathbf{0}$. Equivalently

$$\boldsymbol{v}_1 + \cdots + \boldsymbol{v}_{n-1} = -\boldsymbol{s}.$$

Now we prove that there exists no vector $\boldsymbol{v}$ satisfying that $\boldsymbol{v} \in M_2$ and $\boldsymbol{v} \notin \{\boldsymbol{v}_1, \cdots, \boldsymbol{v}_{n-1}\}$.

If there exists such vector $\boldsymbol{v}$, then $\boldsymbol{v} \cdot \boldsymbol{s} \leqslant 0$. Hence

$$|\boldsymbol{v}_1 + \cdots + \boldsymbol{v}_{n-1} + \boldsymbol{v}|^2 = |\boldsymbol{v} - \boldsymbol{s}|^2 = |\boldsymbol{s}|^2 - 2\boldsymbol{v} \cdot \boldsymbol{s} + |\boldsymbol{v}|^2 > |\boldsymbol{s}|^2,$$

which is contrary to how we choose $\boldsymbol{u}_1, \cdots, \boldsymbol{u}_n$.

Therefore $|M_2| \leqslant n-1$.

On the other hand, it can be deduced from (1) that there exist $\boldsymbol{u}'_1, \cdots, \boldsymbol{u}'_n \in M$, satisfying that

$$|\boldsymbol{u}_1 + \cdots + \boldsymbol{u}_n + \boldsymbol{u}'_1 + \cdots + \boldsymbol{u}'_n| = 0.$$

Equivalently, $\boldsymbol{u}'_1 + \cdots + \boldsymbol{u}'_n = -\boldsymbol{s}$. It can be deduced from (2) that there exist $\boldsymbol{v}'_1, \cdots, \boldsymbol{v}'_{n-1}$, satisfying that $\boldsymbol{u}'_1 + \cdots + \boldsymbol{u}'_n + \boldsymbol{v}'_1 + \cdots + \boldsymbol{v}'_{n-1} = \mathbf{0}$. Equivalently, $\boldsymbol{v}'_1 + \cdots + \boldsymbol{v}'_{n-1} = \boldsymbol{s}$.

Similarly, we can prove that there exists no vector $\boldsymbol{v}' \in M_1$, satisfying that $\boldsymbol{v}' \notin \{\boldsymbol{v}'_1, \cdots, \boldsymbol{v}'_{n-1}\}$. Hence $|M_1| \leqslant n-1$.

It can be then inferred that $|M| \leqslant 2n-2$, which is contradictory. Hence there is no M satisfying such conditions.

Example 6. Let α be a given positive integer. Find the greatest positive integer β satisfying that there exist $x, y \in \mathbf{N}^*$, such that

$$\beta = \frac{x^2 + y^2 + \alpha}{xy}. \quad ①$$

Solution. Noting when $\beta = \alpha + 2$, we can take $x = y = 1$ and then ① holds. Hence $\beta_{\max} \geqslant \alpha + 2$.

On the other hand, let β be a positive integer satisfying the condition. Assume that (x, y) is the one among all the pairs of positive integers satisfying ① (here we regard β as constant) that $x + y$ is least.

If $x = y$, then $\beta = \frac{2x^2 + \alpha}{x^2} = 2 + \frac{\alpha}{x^2} \leqslant 2 + \alpha$.

If $x \neq y$, without loss of generality, assume that $x > y$. Then there exists another real root $\bar{x}$ of the quadratic equation about x

$$x^2 - \beta y \cdot x + y^2 + \alpha = 0. \quad ②$$

It can be inferred from Vieta Theorem and ② that $\bar{x} = \beta y - x \in \mathbf{Z}$. Noting that $x \cdot \bar{x} = y^2 + \alpha$, or $\bar{x} = \dfrac{y^2 + \alpha}{r} > 0$, we have that $\bar{x}$ is a positive integer. Then $(\bar{x}, y)$ is a pair of positive integers that satisfies ① as well. Therefore

$$\bar{x} + y = \frac{y^2 + \alpha}{x} + y \geqslant x + y.$$

The minimality of $x + y$ is made use of here.

Then we have $x^2 \leqslant y^2 + \alpha$ and $x \geqslant y + 1$. Moreover,

$$\begin{aligned}\beta &= \frac{x^2 + y^2 + \alpha}{xy} \leqslant \frac{2(y^2 + \alpha)}{xy} \leqslant \frac{2(y^2 + \alpha)}{y(y+1)} \\ &= \frac{2y^2}{y^2 + y} + \frac{2\alpha}{y(y+1)} < 2 + \alpha.\end{aligned}$$

This shows that $\beta_{\max} \leqslant 2 + \alpha$.

In conclusion, the maximum of β is $\alpha + 2$.

Example 7. Find all integers $n > 1$ that any divisor, greater than 1, of them, can be written in the form of $a^r + 1$, where $a, r \in \mathbf{N}^*$, $r \geqslant 2$.

Solution. Let S be the set of all the positive integers satisfying the condition, i.e., for any $n \in S$, $n > 1$, any divisor, greater than 1, of it, can be written in the form of $a^r + 1$, where $a, r \in \mathbf{N}^*$, $r \geqslant 2$.

It can be deduced that for any $n \in S$ $(n > 2)$, there exist $a, r \in \mathbf{N}^*$, $a, r > 1$, satisfying that $n = a^r + 1$. Suppose that a is the least when n is written in this form, i.e., there is no $b, t \in \mathbf{N}^*$, $t > 1$, satisfying that $a = b^t$. Then r must be even (otherwise, let r be odd. Then $(a+1) \mid n$. Hence $a+1$ can be written in the form of $b^t + 1$. Then $a = b^t$, which is contrary to the minimality of a). Therefore each element greater than 1 in S can be written in the form of $n = x^2 + 1$, $x \in \mathbf{N}^*$.

Now let's find every n in S.

If n is a prime number, then n is a prime number that can be written in the form of $x^2 + 1$.

If n is a composite number, discuss the problem in two cases:

(1) If n is an odd composite number, then there exist odd prime numbers p and q that p, q, $pq \in S$. Then there should be a, b, $c \in \mathbf{N}^*$ satisfying that

$$p = 4a^2 + 1, \ q = 4b^2 + 1, \ pq = 4c^2 + 1.$$

We can assume that $a \leqslant b < c$. Therefore $pq - q = 4(c^2 - b^2)$ and thus, $q \mid 4(c - b)(c + b)$. Noting that q is an odd prime number, we have that $q \mid c - b$ or $q \mid c + b$. Then there must be $q < 2c$ and lead to $pq < 4c^2 < 4c^2 + 1 = pq$, which is contradictory.

(2) If n is an even composite number, noting that $2^2 \notin S$ and combining with the previous discussion, we have that n can only be written in the form of $2q$, where q is an odd prime number. Now q, $2q \in S$. Hence there exist a, $b \in \mathbf{N}^*$ satisfying that

$$q = 4a^2 + 1, \ 2q = b^2 + 1.$$

Hence $q = b^2 - 4a^2 = (b - 2a)(b + 2a)$ and thus, $b - 2a = 1$, $b + 2a = q$. Moreover $q - 1 = 4a$. Noting that $q - 1 = 4a^2$, we have $4a = 4a^2$ and then $a = 1$, $b = 3$, $q = 5$, $n = 10$. That is, there is only one even composite number 10 in S.

In conclusion, any $n \in S$ must be 10 or prime number written in the form of $x^2 + 1$. It is obvious that such n satisfies the conditions in the problem. Hence

$$S = \{x^2 + 1 \mid x \in \mathbf{N}^*, \ x^2 + 1 \text{ is prime}\} \cup \{10\}.$$

Example 8. There are two piles of coins on the table. It is given that the total weight of the two piles are equivalent. Furthermore, for any positive number k (which is less than the number of coins in either pile), the total weight of the heaviest k coins in the first pile do not exceed the total weight of their counterparts in the second pile. For any positive number x, prove that if we replace coins weighing no less than x in both piles by coins weighing exactly x, then the total weight of the first pile will not be less than the one of the second pile, after the operation.

Proof. Let's handle the problem with the axiom of order.

Suppose that the weights of the coins in the first pile are $x_1 \geqslant \cdots \geqslant x_n$ and the weights of the coins in the second pile are $y_1 \geqslant \cdots \geqslant y_m$. It can be deduced from the condition that for any $k \leqslant \min\{m, n\}$, $x_1 + \cdots + x_k \leqslant y_1 + \cdots + y_k$.

For any $x \in \mathbf{R}$, let

$$x_1 \geqslant \cdots \geqslant x_s \geqslant x > x_{s+1} \geqslant \cdots \geqslant x_n,$$
$$y_1 \geqslant \cdots \geqslant y_t \geqslant x > y_{t+1} \geqslant \cdots \geqslant y_m.$$

We need to prove that

$$sx + x_{s+1} + \cdots + x_n \geqslant tx + y_{t+1} + \cdots + y_m. \quad ①$$

It is obvious that when s or t does not exist (note that the condition implies that if t does not exist, s does not exist as well), the inequality ① can be deduced by the equality $x_1 + \cdots + x_n = y_1 + \cdots + y_m$. Hereafter, let's consider the situation where s and t exist.

Let $x_1 + \cdots + x_n = y_1 + \cdots + y_m = A$. Then ① is equivalent to the following inequality.

$$sx + (A - x_1 - \cdots - x_s) \geqslant tx + (A - y_1 - \cdots - y_t)$$
$$\Leftrightarrow x_1 + \cdots + x_s + (t - s)x \leqslant y_1 + \cdots + y_t. \quad ②$$

If $t \geqslant s$, then

$$x_1 + \cdots + x_s + (t - s)x = x_1 + \cdots + x_s + \underbrace{x + \cdots + x}_{t-s}$$
$$\leqslant y_1 + \cdots + y_s + y_{s+1} + \cdots + y_t.$$

Inequality ② holds.

If $t < s$, then ② is equivalent to

$$x_1 + \cdots + x_s \leqslant y_1 + \cdots + y_t + \underbrace{x + \cdots + x}_{s-t}. \quad ③$$

It can be deduced from the condition that

$$\begin{aligned} x_1 + \cdots + x_s &\leqslant y_1 + \cdots + y_t + y_{t+1} + \cdots + y_s \\ &\leqslant y_1 + \cdots + y_t + \underbrace{x + \cdots + x}_{s-t}. \end{aligned}$$

Hence ③ holds.

In conclusion, the proposition is true.

4 General Terms and Summation of Sequences

A row of numbers arranged in a certain order is called sequence. Each number in the sequence is called term. The numbers are called, in order, the first term, the second term, ..., the nth term, ...

The general form of sequences can be written as

$$a_1, a_2, \cdots, a_n, \cdots.$$

It is denoted by $\{a_n\}$. If the nth term a_n of $\{a_n\}$ can be expressed by an algebraic formula, then the formula is called the general formula.

By the definition of sequence above, a sequence is essentially a function defined on the set of positive integers. Finding the general formula and adding up the first n terms of a sequence are most fundamental and common among relevant problems.

Let S_n be the sum of first n terms of $\{a_n\}$. Then its relationship with the general terms is as follows.

$$a_1 = S_1, a_n = S_n - S_{n-1}, n = 2, 3, \cdots.$$

For the sake of convenience of discussion, let's introduce the following concepts.

If there are finite number of terms in a sequence, then we call it a finite sequence. Otherwise, it is called an infinite sequence.

If sequence $\{a_n\}$ satisfies that for any $n \in \mathbf{N}^*$, $a_n < a_{n+1}$ (resp. $a_n > a_{n+1}$), then it is called an increasing (resp. decreasing) sequence. If $a_n \leqslant a_{n+1}$ (resp. $a_n \geqslant a_{n+1}$), then it is called a non-decreasing (resp. non-increasing) sequence.

If there exists a constant M satisfying that for any $n \in \mathbf{N}^*$, $|a_n| \leqslant M$. *Then the real sequence* $\{a_n\}$ is called a bounded sequence.

Example 1. Sequence $\{a_n\}$ satisfies that for any nonnegative integers m, $n(m \geqslant n)$, $a_{m+n} + a_{m-n} = \frac{1}{2}(a_{2m} + a_{2n})$. Besides, $a_1 = 1$. Find the general formula of the sequence.

Solution. It can be inferred from the condition in the problem that for any $m \in \mathbf{N}$ it holds that

$$\frac{1}{2}(a_{2m} + a_{2m}) = a_{2m} + a_0 = 2(a_m + a_m),$$

then $a_0 = 0$ and $a_{2m} = 4a_m$.

Noting that $a_1 = 1$, we can find that when $n \in \{0, 1, 2\}$, it holds that $a_n = n^2$. Then is it the general formula of the sequence?

If $a_{m-1} = (m-1)^2$, $a_m = m^2$, it can be deduced from the condition that

$$a_{m+1} + a_{m-1} = \frac{1}{2}(a_{2m} + a_2) = \frac{1}{2}(4a_m + 4a_1).$$

Hence $a_{m+1} = 2a_m - a_{m-1} + 2a_1 = 2m^2 - (m-1)^2 + 2 = m^2 + 2m + 1 = (m+1)^2$. Then by the principle of mathematical induction, we have that for $n \in \mathbf{N}^*$, $a_n = n^2$.

In conclusion, the general formula of the sequence is $a_n = n^2$.

Explanation. It is common, in sequence questions, that we are asked to find the general formula of a sequence with conditions given. This question is also a problem of functional equation in a special form, since sequences are functions defined on the set of positive integers.

Example 2. Let n be a given positive integer. Sequence a_0, a_1, a_2, $\cdots$, a_n satisfies that $a_0 = \frac{1}{2}$, $a_k = a_{k-1} + \frac{a_{k-1}^2}{n}$, $k = 1, 2, \cdots, n$. Prove that $1 - \frac{1}{n} < a_n < 1$.

Proof. It can be inferred from the condition that for any $1 \leqslant k \leqslant n$, we have $a_{k-1} < a_k$. Hence for any $0 \leqslant k \leqslant n$, $a_k > 0$.

Let's transform the formula in the problem and we can get

$$\frac{1}{a_k} = \frac{n}{na_{k-1} + a_{k-1}^2} = \frac{1}{a_{k-1}} - \frac{1}{a_{k-1} + n}.$$

After transposition of terms, we have

$$\frac{1}{a_{k-1} + n} = \frac{1}{a_{k-1}} - \frac{1}{a_k}. \qquad ①$$

Adding up ① according to the subscript k from 1 to n, we get that

$$\sum_{k=1}^{n} \frac{1}{a_{k-1} + n} = \left(\frac{1}{a_0} - \frac{1}{a_1}\right) + \left(\frac{1}{a_1} - \frac{1}{a_2}\right) + \cdots + \left(\frac{1}{a_{n-1}} - \frac{1}{a_n}\right)$$
$$= \frac{1}{a_0} - \frac{1}{a_n} = 2 - \frac{1}{a_n}.$$

Noting that $a_{k-1} > 0$, we have that

$$2 - \frac{1}{a_n} = \sum_{k=1}^{n} \frac{1}{a_{k-1} + n} < \sum_{k=1}^{n} \frac{1}{n} = 1.$$

Hence $a_n < 1$.

Since $a_k > a_{k-1}$, we get $0 < a_0 < a_1 < \cdots < a_n < 1$. Therefore

$$2 - \frac{1}{a_n} = \sum_{k=1}^{n} \frac{1}{a_{k-1} + n} > \sum_{k=1}^{n} \frac{1}{1 + n} = \frac{n}{n + 1}.$$

Thus,

$$a_n > \frac{n + 1}{n + 2} = 1 - \frac{1}{n + 2} > 1 - \frac{1}{n}.$$

Hence the proposition is true.

Explanation. The skill of taking the reciprocal of both sides of the formula is innovated by the idea of splitting terms, which is often applied in the summation of sequences so that we can cancel out the former and latter terms.

Example 3. For $n \in \mathbf{N}^*$, let $a_n = \dfrac{n}{(n-1)^{\frac{4}{3}} + n^{\frac{4}{3}} + (n+1)^{\frac{4}{3}}}$. Prove that

$$a_1 + a_2 + \cdots + a_{999} < 50.$$

Proof. The general idea is to deal with the problem "from part to the whole". For this purpose, we will enlarge a_n properly. We can thus, split the terms and cancel them out.

Noting that $x^3 - y^3 = (x-y)(x^2+xy+y^2)$, let $x = (n+1)^{\frac{2}{3}}$, $y = (n-1)^{\frac{2}{3}}$. Then $xy = (n^2-1)^{\frac{2}{3}} < n^{\frac{4}{3}}$. Hence

$$\begin{aligned} a_n &< \frac{n}{x^2+xy+y^2} = \frac{n(x-y)}{x^3-y^3} \\ &= \frac{n(x-y)}{(n+1)^2-(n-1)^2} = \frac{1}{4}(x-y) \\ &= \frac{1}{4}\left((n+1)^{\frac{2}{3}} - (n-1)^{\frac{2}{3}}\right). \end{aligned}$$

Therefore

$$\begin{aligned} a_1 + \cdots + a_{999} &< \frac{1}{4}\sum_{n=1}^{999}\left((n+1)^{\frac{2}{3}} - (n-1)^{\frac{2}{3}}\right) \\ &= \frac{1}{4}\left(\sum_{n=2}^{1000} n^{\frac{2}{3}} - \sum_{n=0}^{998} n^{\frac{2}{3}}\right) \\ &= \frac{1}{4}(1000^{\frac{2}{3}} + 999^{\frac{2}{3}} - 1) \\ &< \frac{1}{2} \times 1000^{\frac{2}{3}} = 50. \end{aligned}$$

The proposition is thus, proved.

Explanation. It is an important method to lessen or enlarge the terms first and then add them up, when dealing with inequalities relevant to the summation of sequences.

Example 4. Let $k \in \mathbf{N}^*$ and $k \equiv 3 \pmod 4$. Define that

$$S_n = C_n^0 - C_n^2 k + C_n^4 k^2 - C_n^6 k^3 + \cdots.$$

For any $n \in \mathbf{N}^*$, prove that $2^{n-1} \mid S_n$.

Proof. Let's prove by using complex numbers.

If can be deduced from the definition of S_n and the binomial theorem that

$$S_n = \mathrm{Re}(1 + \sqrt{k}\,\mathrm{i})^n,$$

where i is the imaginary unit. $\mathrm{Re}(z)$ denotes the real part of z.

Hence

$$S_n = \frac{1}{2}((1 + \sqrt{k}\,\mathrm{i})^n + (1 - \sqrt{k}\,\mathrm{i})^n).$$

Moreover letting $x = 1 + \sqrt{k}\,\mathrm{i}$, $y = 1 - \sqrt{k}\,\mathrm{i}$, we have

$$\begin{aligned} S_{n+2} &= \frac{1}{2}(x^{n+2} + y^{n+2}) \\ &= \frac{1}{2}((x^{n+1} + y^{n+1})(x + y) - xy(x^n + y^n)) \\ &= (x + y)S_{n+1} - xyS_n \\ &= 2S_{n+1} - (1 + k)S_n. \end{aligned}$$

Besides, $S_1 = 1$, $S_2 = 1 - k$.

Now we prove that for any $n \in \mathbf{N}^*$, $2^{n-1} \mid S_n$.

Noting that $k \equiv 3 \pmod 4$, we have that the proposition is true for $n = 1, 2$. Then suppose that the proposition is true for n, $n + 1$. That is, $2^{n-1} \mid S_n$ and $2^n \mid S_{n+1}$. As to the case of $n + 2$, since $1 + k \equiv 1 + 3 \equiv 0 \pmod 4$ and

$$S_{n+2} = 2S_{n+1} - (1 + k)S_n,$$

we have that $2^{n+1} \mid S_{n+2}$ (because $2^{n+1} \mid 2S_{n+1}$, $2^{n+1} \mid (1 + k)S_n$). Therefore for any $n \in \mathbf{N}^*$, $2^{n-1} \mid S_n$.

Explanation. We began from the conditions, then made proper transformations, established the recurrence relation and finally made use of mathematical induction. Thus, the problem was accomplished by us at a stretch with clear thinking. This idea is also quite easy to realize.

Example 5. The first several terms of increasing sequence of positive integers $\{a_n\}$ are

$$1, 2, 4, 5, 7, 9, 10, 12, 14, 16, 17, \cdots$$

The structure is one odd number, two even numbers, three odd numbers, four even numbers, ...

For any $n \in \mathbf{N}^*$, prove that $a_n = 2n - \left[\frac{1+\sqrt{8n-7}}{2}\right]$.

Proof. If there exists $k \in \mathbf{N}^*$, satisfying that $n = 1+2+\cdots+k$, then n is called a triangle number.

Now we define the sequence $\{b_n\}$: $b_1 = 1$,

$$b_{n+1} - b_n = \begin{cases} 1, & \text{if } n \text{ is a triangle number,} \\ 2, & \text{if } n \text{ is not a triangle number.} \end{cases} \qquad ①$$

Then it can be inferred from the structure of $\{a_n\}$ and mathematical induction that $a_n = b_n$.

Moreover, since sequence $\{b_n\}$ satisfying ① exists uniquely, it suffices to prove that $\left(\text{note that when } n = 1,\ 2n - \left[\frac{1+\sqrt{8n-7}}{2}\right] = 1\right)$

$$c_n = \begin{cases} 1, & \text{if } n \text{ is a triangle number,} \\ 2, & \text{if } n \text{ is not a triangle number.} \end{cases} \qquad ②$$

Here

$$c_n = 2(n+1) - \left[\frac{1+\sqrt{8(n+1)-7}}{2}\right] - \left(2n - \left[\frac{1+\sqrt{8n-7}}{2}\right]\right).$$

For this purpose, let's prove:

If and only if n is a triangle number,

$$\frac{1+\sqrt{8(n+1)-7}}{2} \in \mathbf{N}^*. \qquad ③$$

In fact, if there exists $k \in \mathbf{N}^*$ satisfying that

$$n = 1+2+\cdots+k = \frac{k(k+1)}{2},$$

then

$$\frac{1+\sqrt{8(n+1)-7}}{2}=\frac{1+\sqrt{4k(k+1)+1}}{2}$$
$$=\frac{1+2k+1}{2}=k+1\in\mathbf{N}^*.$$

On the other hand, if n is not a triangle number, there exists $k\in\mathbf{N}^*$, satisfying that $\frac{k(k+1)}{2}<n<\frac{(k+1)(k+2)}{2}$ (i.e., n is between two consecutive triangle numbers). Calculating in the same way, we get that

$$k+1<\frac{1+\sqrt{8(n+1)-7}}{2}<k+2.$$

Hence ③ holds.

Then let's prove that ② holds. Since

$$c_n=2+\left[\frac{1+\sqrt{8n-7}}{2}\right]-\left[\frac{1+\sqrt{8(n+1)-7}}{2}\right],$$

when $n\in\mathbf{N}^*$, we have

$$0<\frac{1+\sqrt{8(n+1)-7}}{2}-\frac{1+\sqrt{8n-7}}{2}$$
$$=\frac{1}{2}\left(\sqrt{8n+1}-\sqrt{8n-7}\right)$$
$$=\frac{1}{2}\cdot\frac{8n+1-(8n-7)}{\sqrt{8n+1}+\sqrt{8n-7}}=\frac{4}{\sqrt{8n+1}+\sqrt{8n-7}}$$
$$\leqslant\frac{4}{\sqrt{8+1}+\sqrt{8-7}}=1.$$

Therefore if and only if $\frac{1+\sqrt{8(n+1)-7}}{2}\in\mathbf{N}^*$, $c_n=2-1=1$. For other n, $c_n=2-0=2$.

It can be deduced from ③ that ② holds.

In conclusion, $a_n=2n-\left[\frac{1+\sqrt{8n-7}}{2}\right]$.

Explanation. This is a problem of finding the general formula of a grouping sequence. ① shows the relationship between consecutive terms. The method we applied here is a special one when proving the relationship with the answer already known. It is even harder if we are asked to find the general formula of the sequence by ourselves.

Example 6. A finite sequence $a_0, a_1, \cdots, a_n$ is called k balanced, if

$$\begin{aligned} a_0 + a_k + a_{2k} + \cdots &= a_1 + a_{k+1} + a_{2k+1} + \cdots \\ &= \cdots = a_{k-1} + a_{2k-1} + a_{3k-1} + \cdots \end{aligned}$$

It is given that sequence $a_0, a_1, \cdots, a_{49}$ is k balanced for $k = 3, 5, 7, 11, 13, 17$. Prove that $a_0 = a_1 = \cdots = a_{49} = 0$.

Proof. Consider the polynomial

$$f(x) = a_0 + a_1 x + \cdots + a_{49} x^{49}. \quad ①$$

The idea is to prove that there are 50 distinct complex roots of $f(x)$ and thus, $f(x)$ is a zero polynomial then $a_0 = \cdots = a_{49} = 0$.

For $k \in \{3, 5, 7, 11, 13, 17\}$ let $\varepsilon (\neq 1)$ be a kth root of unity. Then when $m \equiv n \pmod k$, we have $\varepsilon^m = \varepsilon^n$. Hence

$$\begin{aligned} f(\varepsilon) &= (a_0 + a_k + a_{2k} + \cdots) + (a_1 + a_{k+1} + \cdots)\varepsilon + \cdots \\ &\quad + (a_{k-1} + a_{2k-1} + \cdots)\varepsilon^{k-1} \\ &= (a_0 + a_k + a_{2k} + \cdots)(1 + \varepsilon + \varepsilon^2 + \cdots + \varepsilon^{k-1}) \\ &= 0. \end{aligned}$$

We made use of the formula in the condition and that ε is a root of the polynomial $1 + x + \cdots + x^{k-1}$.

Therefore for $k \in \{3, 5, 7, 11, 13, 17\}$ and $\varepsilon = e^{i\frac{2m\pi}{k}} (1 \leqslant m \leqslant k - 1)$, it can be deduced that ε is a complex root of ①. As k varies among different prime numbers, we will get different complex roots. Then there are $(3-1) + (5-1) + \cdots + (17-1) = 50$ roots of $f(x)$. Hence it must be zero polynomial.

The proposition is proved.

Explanation. The polynomial in ① is called the generating function of $\{a_m\}$. This method is often applied in finding the general formula of a sequence. We will refer to generating functions in Section 7 later.

5 Arithmetic Sequences and Geometric Sequences

Arithmetic sequences and geometric sequences are two kinds of the simplest sequences. We often transform other sequences into them.

If the difference (resp. ratio) between the consecutive terms is constant, then the sequence is called an arithmetic (resp. geometric) sequence. The constant is called the common difference (resp. ratio) of the sequence, which is often denoted by d (resp. q). Note that q can't be zero, since zero can't be a denominator.

We have the following formulas relevant to the general formulas and summation of arithmetic sequences and geometric sequences:

1. Let S_n be the sum of the first n terms of $\{a_n\}$. Then

$$a_n = a_1 + (n-1)d,$$
$$S_n = \frac{1}{2}(a_1 + a_n)n = a_1 n + \frac{n(n-1)}{2}d.$$

2. Let S_n be the sum of the first n terms of $\{a_n\}$. Then $a_n = a_1 \cdot q^{n-1}$ and

$$S_n = \begin{cases} na_1, & \text{if } q = 1, \\ \dfrac{a_1(1-q^n)}{1-q}, & \text{if } q \neq 1. \end{cases}$$

3. If geometric sequence $\{a_n\}$ is an infinite sequence and its common ratio q satisfies that $|q| < 1$, then it is called an infinite decaying geometric sequence. The sum of all of its terms is $S = \frac{a_1}{1-q}$.

Example 1. Let's arrange $n^2 (n \geqslant 4)$ positive real numbers in n lines and n rows:

$$\begin{matrix} a_{11} & a_{12} & \cdots & a_{1n} \\ a_{21} & a_{22} & \cdots & a_{2n} \\ \cdots & \cdots & \cdots & \cdots \\ a_{n1} & a_{n2} & \cdots & a_{nn} \end{matrix}$$

The numbers in each line constitute an arithmetic sequence, while the numbers in each row constitute a geometric sequence. Furthermore, the common ratios are the same. Given that $a_{24} = 1$, $a_{42} = \frac{1}{8}$, $a_{43} = \frac{3}{16}$. Find the value of $a_{11} + a_{22} + \cdots + a_{nn}$.

Solution. Let the common ratio of the geometric sequence constituted by numbers in each row be q. Then $a_{44} = a_{24} \cdot q^2 = q^2$.

Since the numbers in the fourth line constitute an arithmetic sequence, a_{42}, a_{43}, a_{44} constitute an arithmetic sequence as well. Hence $a_{42} + a_{44} = 2a_{43}$. Then

$$\frac{1}{8} + q^2 = \frac{6}{16}.$$

Therefore $q^2 = \frac{1}{4}$. Noting that all of the numbers in the table are real positive numbers, we have that $q = \frac{1}{2}$.

Since the numbers in the fourth line constitute an arithmetic sequence and $a_{42} = \frac{1}{8}$, $a_{43} = \frac{3}{16}$, it can be deduced that the common difference of the arithmetic sequence is $\frac{3}{16} - \frac{1}{8} = \frac{1}{16}$. Hence the first term $a_{41} = \frac{1}{8} - \frac{1}{16} = \frac{1}{16}$. Then for any $1 \leqslant k \leqslant n$, $a_{4k} = \frac{k}{16}$.

Noting that the kth row is a geometric sequence whose common ratio is $\frac{1}{4}$, we have that

$$a_{1k} = a_{4k} \cdot q^{-3} = \frac{k}{16} \cdot \left(\frac{1}{2}\right)^{-3} = \frac{k}{2}.$$

Therefore for any $1 \leqslant m \leqslant n$, $a_{mk} = a_{1k} \cdot q^{m-1} = \frac{k}{2^m}$. Moreover

$$a_{mm} = \frac{m}{2^m}.$$

Let $S = a_{11} + a_{22} + \cdots + a_{nn}$. Then $S = \sum_{m=1}^{n} \frac{m}{2^m}$ and thus, $\frac{S}{2} = \sum_{m=1}^{n} \frac{m}{2^{m+1}}$. Subtracting both sides of the two formulas, we get

$$\begin{aligned}
\frac{1}{2}S &= \sum_{m=1}^{n} \frac{m}{2^m} - \sum_{m=1}^{n} \frac{m}{2^{m+1}} \\
&= \sum_{m=1}^{n} \frac{m}{2^m} - \sum_{m=2}^{n+1} \frac{m-1}{2^m} \\
&= \frac{1}{2} + \sum_{m=2}^{n} \left(\frac{m}{2^m} - \frac{m-1}{2^m} \right) - \frac{n}{2^{n+1}} \\
&= \frac{1}{2} + \sum_{m=2}^{n} \frac{1}{2^m} - \frac{n}{2^{n+1}} \\
&= \sum_{m=1}^{n} \frac{1}{2^m} - \frac{n}{2^{n+1}} = 1 - \frac{1}{2^n} - \frac{n}{2^{n+1}} \\
&= 1 - \frac{(n+2)}{2^{n+1}}.
\end{aligned}$$

Hence $S = 2 - \frac{n+2}{2^n}$.

Example 2. Given that the nonnegative real solutions to the equation about x

$$(2a-1)\sin x + (2-a)\sin 2x = \sin 3x$$

constitutes an infinite arithmetic sequence, in ascending order. Find the range of a.

Solution. The equation can be transformed into

$$\begin{aligned}
& 2a\sin x - a\sin 2x + 2\sin 2x - \sin x - \sin 3x = 0 \\
\Leftrightarrow\ & 2a\sin x(1-\cos x) + 2\sin 2x - 2\sin 2x\cos x = 0 \\
\Leftrightarrow\ & (2a\sin x - 2\sin 2x)(1-\cos x) = 0.
\end{aligned}$$

Then there holds $1 - \cos x = 0$ or $\sin 2x = a \sin x$.

The nonnegative real solutions to the former one are $x = 2k_1\pi$, $k_1 \in \mathbf{N}$. As to the latter, it can be transformed into $\sin x = 0$ or $\cos x = \frac{a}{2}$. The nonnegative real solutions of $\sin x = 0$ are $x = k_2\pi$, $k_2 \in \mathbf{N}$. There exist solutions to $\cos x = \frac{a}{2}$ if and only if $|a| \leqslant 2$ and the nonnegative real solutions are $x = 2k_3\pi + \arccos\frac{a}{2}$ or $x = 2k_4\pi + \pi + \arccos\frac{a}{2}$, k_3, $k_4 \in \mathbf{N}$.

In conclusion, when $|a| \geqslant 2$, the nonnegative real solutions to the equation are $x = k\pi$, $k \in \mathbf{N}$, which constitute an arithmetic sequence. When $|a| < 2$, the nonnegative real solutions to the equation are $x = k\pi$, $k \in \mathbf{N}$ or $x = 2k_3\pi + \arccos\frac{a}{2}$ or $x = 2k_4\pi + \pi + \arccos\frac{a}{2}$. Then if and only if $\arccos\frac{a}{2} = \frac{\pi}{2}$, or equivalently $a = 0$, all the nonnegative real solutions to the equation constitute an infinite arithmetic sequence, from less to greater.

Hence the range of a satisfying the condition is

$$a \in (-\infty, -2] \cup \{0\} \cup [2, +\infty).$$

Example 3. There are two infinite sequences of positive integers. One is an arithmetic sequence whose common difference is $d(>0)$, while the other is a geometric sequence whose common ratio is $q(>1)$, where d and q are relatively prime. If there is one common term in both sequences, prove that then there are infinitely many terms in common.

Proof. Let the two sequences are $\{a + nd\}$, $n = 0, 1, 2, \cdots$, and $\{bq^m\}$, $m = 0, 1, 2, \cdots$, respectively, where a, b, d, q are positive integers and $q > 1$.

If there is one term in common, without loss of generality, we assume the first terms of both sequences are same. Otherwise, we can remove the finite number of terms before the common term in both sequences, i.e., $a = b$. Then to prove that there are infinitely many terms in common, we need only to prove that there exist infinitely many $m \in \mathbf{N}^*$ satisfying that

$$aq^m \equiv a \pmod d.$$

We need only $q^m \equiv 1 \pmod d$.

Noting that the remainder after dividing $1, q, q^2, \cdots, q^d$ by d ranges among only d numbers, by pigeonhole principle, we have that there exist $0 \leqslant i < j \leqslant d$ satisfying that $q^j \equiv q^i \pmod d$. Since $(d, q) = 1$, it can be deduced that $q^{j-i} \equiv 1 \pmod d$. Moreover for any $n \in \mathbf{N}^*$, letting $m = (j-i)n$. we have that $q^m = (q^{j-i})^n \equiv 1^n = 1 \pmod d$.

Hence the proposition is true.

Explanation. If readers are familiar with Euler's Theorem, we can also find m satisfying the condition, by the fact that $q^{\varphi(d)} \equiv 1 \pmod d$, when $(d, q) = 1$.

Example 4. Sequence $\{a_n\}$ is defined as follows.

$$a_1 = 1\,000\,000, \; a_{n+1} = n\left[\frac{a_n}{n}\right] + n, \; n = 1, 2, \cdots.$$

Prove that there is an infinite subsequence of $\{a_n\}$ which constitutes an arithmetic sequence (the sequence constituted by the terms of another sequence is called a subsequence of it).

Proof. Let $x_n = \frac{a_{n+1}}{n}$. Then for any $n \in \mathbf{N}^*$, $x_n = \left[\frac{a_n}{n}\right] + 1 \in \mathbf{N}^*$. $\{x_n\}$ is a sequence of positive integers.

Moreover for $n \in \mathbf{N}^*$,

$$\begin{aligned} x_{n+1} &= \left[\frac{a_{n+1}}{n+1}\right] + 1 = \left[\frac{nx_n}{n+1}\right] + 1 \\ &= x_n + \left[-\frac{x_n}{n+1}\right] + 1 \\ &\leqslant x_n + (-1) + 1 = x_n. \end{aligned}$$

This shows that $\{x_n\}$ is a non-increasing sequence. Therefore $\{x_n\}$ becomes a constant sequence from some term on (since all the x_n are positive integer). Denote the constant by k. Then from that term on $a_n = kn$. Hence $\{a_n\}$ is an arithmetic sequence from that term on.

The proposition is proved.

Explanation. The conclusion in this problem does not rely on the initial value (we only need that $a_1 \geqslant 0$). We applied an obvious result in solving this problem that any infinite non-increasing sequence of positive integers becomes constant from some term on.

Example 5. For any given positive integer $n \geqslant 3$. Prove that there exists an arithmetic sequence $a_1, a_2, \cdots, a_n$ and a geometric sequence $b_1, b_2, \cdots, b_n$ satisfying that

$$b_1 < a_1 < b_2 < a_2 < \cdots < b_n < a_n. \qquad ①$$

Proof. Note that exponential growth is greater than linear growth. Hence there doesn't exist an infinite increasing arithmetic sequence of positive integers $\{a_m\}$ and an infinite increasing geometric sequence of positive integers $\{b_m\}$ satisfying that for any $m \in \mathbf{N}^*$, $a_m > b_m$, letting alone satisfying ①. What we discuss in this problem is about finite sequences. The idea is to let the common ratio be approximately 1 and keep enough space between consecutive terms.

Consider the sequences $\{a_n\}$ and $\{b_n\}$ defined by the following formula.

$$b_1 = x^n,\ b_2 = x^{n-1}(1+x),\ \cdots,\ b_n = x(1+x)^{n-1};$$
$$a_m = x^{n-1}(1+x) - 1 + (m-1)x^{n-1},\ m = 1, 2, \cdots, n.$$

Here x is an undetermined positive integer. Then $\{a_m\}$ is an arithmetic sequence whose common difference is x^{n-1}, while $\{b_m\}$ is a geometric sequence whose common ratio is $1+\dfrac{1}{x}$. Hence we need only prove that there exists such positive integer that ① holds.

On one hand, for $1 \leqslant m \leqslant n$, since

$$a_m = x^n + x^{n-1} - 1 + (m-1)x^{n-1},$$

we have $a_m > x^n$, when $x > 1$. Hence

$$a_{m+1} = a_m + x^{n-1} < a_m + \frac{a_m}{x} = a_m\left(1 + \frac{1}{x}\right).$$

Noting that $a_1 = b_2 - 1 < b_2$ and that $\{b_n\}$ is a geometric sequence whose common ratio is $1 + \dfrac{1}{x}$, it can be deduced that for $1 \leqslant m \leqslant n - 1$, $a_m < b_{m+1}$.

On the other hand, let's prove that there exists $x \in \mathbf{N}^*$ $(x > 1)$, satisfying that for any $1 \leqslant m \leqslant n$, $b_m < a_m$.

In fact,

$$\begin{aligned} b_m < a_m &\Leftrightarrow x^{n-m+1}(1+x)^{m-1} < x^n + mx^{n-1} - 1 \\ &\Leftrightarrow x^n + C_{m-1}^{m-2}x^{n-1} + C_{m-1}^{m-3}x^{n-2} + \cdots + C_{m-1}^{0}x^{n-m+1} < x^n + mx^{n-1} - 1 \\ &\Leftrightarrow C_{m-1}^{m-3}x^{n-2} + \cdots + C_{m-1}^{0}x^{n-m+1} < x^{n-1} - 1 \\ &\Leftrightarrow x(C_{m-1}^{m-3}x^{n-3} + \cdots + C_{m-1}^{0}x^{n-m}) < x^{n-1} - 1. \qquad ② \end{aligned}$$

Since $n \geqslant m$, it can be deduced that

$$C_{n-1}^{m-3}x^{n-3} + \cdots + C_{n-1}^{m-1}x^{n-m} + C_{n-1}^{1}x + C_{n-1}^{0} \geqslant C_{m-1}^{m-3}x^{n-3} + \cdots + C_{m-1}^{m-1}x^{n-m}.$$

Hence if

$$x(C_{n-1}^{n-3}x^{n-3} + \cdots + C_{n-1}^{1}x + C_{n-1}^{0}) < x^{n-1} - 1 \qquad ③$$

holds, then ② holds.

The left side of ③ is a polynomial of degree $n-2$ of x, while the right side is a polynomial of degree $n-1$ of x. Therefore when x is sufficiently big, ③ holds.

In conclusion, the sequence satisfying the conditions exists.

Example 6. Let $k\,(\geqslant 2)$ be a given positive integer. For any $1 \leqslant i \leqslant k$, a_i and d_i are positive integers. The set corresponding to the arithmetic sequence $\{a_i + nd_i\}$ $(n = 0, 1, 2, \cdots)$ is $A_i = \{a_i + nd_i \mid n = 0, 1, 2, \cdots\}$, $1 \leqslant i \leqslant k$. Let $A_1, A_2, \cdots, A_k$ be a k-partition of $\mathbf{N}^*$ (i.e., the intersection of any two of $A_1, A_2, \cdots, A_k$ is empty and $A_1 \cup \cdots \cup$

$A_k = \mathbf{N}^*$). Prove that

(1) $\frac{1}{d_1} + \cdots + \frac{1}{d_k} = 1$;

(2) $\frac{a_1}{d_1} + \cdots + \frac{a_k}{d_k} = \frac{k+1}{2}$.

Proof. Let's make use of the method of generating functions. It can be inferred from the conditions in the problem that for $|x| < 1$,

$$\sum_{m=1}^{+\infty} x^m = \sum_{i=1}^{k} \left(\sum_{n=0}^{+\infty} x^{a_i + nd_i} \right).$$

By the summation formula of the infinite decaying geometric sequence, we have

$$\frac{x}{1-x} = \sum_{i=1}^{k} \frac{x^{a_i}}{1 - x^{d_i}}.$$

Hence

$$x = \sum_{i=1}^{k} \frac{x^{a_i}}{1 + x + \cdots + x^{d_i - 1}}. \qquad ①$$

As x approaches 1 from the left, take the limit of both sides. It yields $\sum_{i=1}^{k} \frac{1}{d_i} = 1$. Hence (1) holds.

Now differentiate both sides of ① with respect to x. We have

$$1 = \sum_{i=1}^{k} \frac{a_i x^{a_i - 1}(1 + x + \cdots + x^{d_i - 1}) - x^{a_i}(0 + 1 + 2x + \cdots + (d_i - 1)x^{d_i - 2})}{(1 + x + \cdots + x^{d_i - 1})^2}.$$

Then as x approaches 1 from the left, take the limit. It yields

$$1 = \sum_{i=1}^{k} \frac{a_i d_i - (1 + 2 + \cdots + (d_i - 1))}{d_i^2}.$$

Hence

$$\sum_{i=1}^{k} \frac{a_i}{d_i} = 1 + \sum_{i=1}^{k} \frac{(d_i - 1)d_i}{2d_i^2}$$
$$= 1 + \frac{1}{2} \sum_{i=1}^{k} \left(1 - \frac{1}{d_i}\right)$$

$$= 1 + \frac{1}{2}(k - 1) = \frac{k + 1}{2}.$$

We applied (1) here.

Therefore the proposition is true.

Explanation. In comparison to Example 6 in last section, we referred to the theory of infinite series, an important skill in utilizing the generating function. We will discuss in detail what generating functions are and how to utilize them in Section 7.

6 Higher-order Arithmetic Sequences and the Method of Differences

We can get a new sequence by subtracting consecutive terms of a given sequence $\{a_n\}$:

$$a_2 - a_1, a_3 - a_2, \cdots, a_{n+1} - a_n, \cdots.$$

It is called the first difference sequence of $\{a_n\}$. If this sequence is denoted by $\{b_n\}$, where $b_n = a_{n+1} - a_n$. Then subtracting the consecutive terms of $\{b_n\}$, we get the following sequence.

$$b_2 - b_1, b_3 - b_2, \cdots, b_{n+1} - b_n, \cdots.$$

It is called the second difference sequence of $\{a_n\}$.

And similarly, for any $p \in \mathbf{N}^*$, we can define the pth difference sequence of $\{a_n\}$.

If the pth difference sequence of $\{a_n\}$ is a nonzero constant sequence, then $\{a_n\}$ is called a pth order arithmetic sequence. Specifically, the first order arithmetic sequence is the arithmetic sequence in general. The second and higher order ones are called higher-order arithmetic sequences in general.

Note that sequences are functions defined on $\mathbf{N}^*$. Generalizing the idea of subtraction above, we can get the concept of difference.

Suppose $f(x)$ is a function defined on $\mathbf{R}$. Let $\Delta f(x) = f(x+1) - f(x)$. Then $\Delta f(x)$ is also a function defined on $\mathbf{R}$. It is called the first difference of $f(x)$. Similarly, we can recursively define the second, the third, ..., the pth difference of $f(x)$.

$$\begin{aligned}\Delta^2 f(x) &= \Delta(\Delta f(x)) = \Delta(f(x+1) - f(x))\\ &= (f(x+2) - f(x+1)) - (f(x+1) - f(x))\\ &= f(x+2) - 2f(x+1) + f(x),\\ &\cdots,\\ \Delta^p f(x) &= \Delta(\Delta^{p-1} f(x)).\end{aligned}$$

By mathematical induction, we can prove the following theorem.

Theorem 1. Let $f(x)$ be a function defined on $\mathbf{R}$. Then

$$\begin{aligned}\Delta^p f(x) &= \sum_{i=0}^{p} (-1)^{p-i} C_p^i f(x+i)\\ &= \sum_{i=0}^{p} (-1)^i C_p^i f(x+p-i).\end{aligned}$$

If function $f(x)$ $(x \in \mathbf{R})$ is a polynomial of degree p of x, then $\Delta f(x)$ is a polynomial of degree $p-1$ of x. $\Delta^2 f(x)$ is a polynomial of degree $p - 2$ of x, $\cdots$, $\Delta^p f(x)$ is a polynomial of degree 0 of x. Besides, $\Delta^p f(x) = p!a_p$, where a_p is the leading coefficient of $f(x)$. When $m > p$, $m \in \mathbf{N}^*$, $\Delta^m f(x) \equiv 0$.

Conversely, for function $f(x)(x \in \mathbf{R})$, if $\Delta^{p+1} f(x) \equiv 0$. then the degree of $f(x)$ is no more than p.

Applying these results to higher-order arithmetic sequences, we have the following theorem.

Theorem 2. The sequence $\{a_n\}$ is a pth order arithmetic sequence if and only if the general term a_n is a polynomial of degree p.

Example 1. Let sequence $\{a_n\}$ be a third order arithmetic sequence. The first several terms are 1, 2, 8, 22, 47, 86, $\cdots$. Find the general formula of $\{a_n\}$.

Solution 1. Calculate the difference sequence of each order of $\{a_n\}$. It yields

$$\{b_n\}:\ 1,\ 6,\ 14,\ 25,\ 39,\cdots;$$
$$\{c_n\}:\ 5,\ 8,\ 11,\ 14,\cdots;$$
$$\{d_n\}:\ 3,\ 3,\ \cdots.$$

Noting that $\{a_n\}$ is a third order arithmetic sequence, we can deduce that $\{d_n\}$ is constant. Then $c_n = c_1 + 3(n-1) = 3n + 2$. Hence

$$b_{n+1} - b_n = 3n + 2,\ n = 1,\ 2,\ \cdots.$$

Therefore

$$\begin{aligned} b_n - b_1 &= (b_n - b_{n-1}) + \cdots + (b_2 - b_1) \\ &= \sum_{k=1}^{n-1}(3k+2) = \frac{3n(n-1)}{2} + 2(n-1) \\ &= \frac{(3n+4)(n-1)}{2}. \end{aligned}$$

So $b_n = \frac{3}{2}n^2 + \frac{1}{2}n - 1$.

The same as above, we have

$$\begin{aligned} a_n - a_1 &= \sum_{k=1}^{n-1}\left(\frac{3}{2}k^2 + \frac{1}{2}k - 1\right) \\ &= \frac{(n-1)n(2n-1)}{4} + \frac{n(n-1)}{4} - (n-1). \end{aligned}$$

It yields $a_n = \frac{1}{2}n^3 - \frac{1}{2}n^2 - n + 2$.

Explanation. Here we applied the method of summing after splitting the terms and the summing formula

$$\sum_{k=1}^{m} k = \frac{m(m+1)}{2},\ \sum_{k=1}^{m} k^2 = \frac{1}{6}m(m+1)(2m+1).$$

Solution 2. By the result of Theorem 2, we can suppose $a_n = An^3 + Bn^2 + Cn + D$, where A, B, C, D are undetermined.

From the initial data, we can deduce that

$$\begin{cases} A + B + C + D = 1, \\ 8A + 4B + 2C + D = 2, \\ 27A + 9B + 3C + D = 8, \\ 64A + 16B + 4C + D = 22. \end{cases}$$

It yields $A = \frac{1}{2}$, $B = -\frac{1}{2}$, $C = -1$, $D = 2$.

Hence $u_n = \frac{1}{2}n^3 - \frac{1}{2}n^2 - n + 2$.

Explanation. The method of undetermined coefficients is often applied in finding the general formula of a higher-order arithmetic sequence.

Example 2. If the value of polynomial $f(x)$ is an integer for any $x \in \mathbf{Z}$, then $f(x)$ is called an integer valued polynomial. For any integer valued polynomial of degree n, prove that there exist integers a_n, a_{n-1}, $\cdots$, a_0, satisfying that

$$f(x) = a_n\binom{x}{n} + a_{n-1}\binom{x}{n-1} + \cdots + a_1\binom{x}{1} + a_0.$$

Here $\binom{x}{k} = \frac{1}{k!}x(x-1)\cdots(x-k+1)$, where $\binom{x}{0} = 1$. It is called a difference polynomial of degree k.

Proof. For polynomial $f(x)$ of degree n, if the leading coefficient is c_n, then letting $b_n = n! \cdot c_n$, we can deduce that $f(x) - b_n\binom{x}{n}$, is a polynomial whose degree $\leqslant n - 1$. Continuing the same process, we have that there exist b_n, b_{n-1}, $\cdots$, $b_0 \in \mathbf{C}$ satisfying that

$$f(x) = b_n\binom{x}{n} + b_{n-1}\binom{x}{n-1} + \cdots + b_1\binom{x}{1} + b_0. \quad ①$$

To prove that the proposition is true we need only to prove that b_n, $\cdots$, b_0 are all integers.

Noting that for $k \in \mathbf{N}^*$,

$$\begin{aligned}\Delta\binom{x}{k} &= \binom{x+1}{k} - \binom{x}{k} \\ &= \frac{1}{k!}((x+1)\cdots(x-k+2) - x(x-1)\cdots(x-k+1)) \\ &= \frac{1}{(k-1)!}x(x-1)\cdots(x-k+2) = \binom{x}{k-1}.\end{aligned}$$

Then it can be deduced from ① that $b_0 = f(0) \in \mathbf{Z}$ (for $f(x)$ is an integer valued polynomial). Take the difference of the both sides of ①. It yields

$$\Delta f(x) = b_n \binom{x}{n-1} + \cdots + b_2 \binom{x}{1} + b_1.$$

Letting $x = 0$, we get $b_1 \in \mathbf{Z}$, and similarly $b_0, b_1, \cdots, b_n$ are all integers.

Explanation. If the value of $\Delta^k f(x)$ at $x = 0$ is denoted by $\Delta^k f(0)$, then from the process of proving that b_k is an integer, it can be deduced that for any polynomial $f(x)$ of degree n,

$$f(x) = \sum_{k=0}^{n} \Delta^k f(0) \binom{x}{k},$$

where $\Delta^0 f(0) = f(0)$.

Example 3. Let sequence $\{a_n\}$ be a pth order arithmetic sequence, whose general formula is $a_n = f(n)$, where $f(x)$ is a polynomial of degree p. Prove that

$$\sum_{m=1}^{n} a_m = \sum_{k=0}^{p} \mathrm{C}_{n+1}^{k+1} \Delta^k f(0). \qquad ①$$

And by this, find the formula of $\sum_{m=1}^{n} m^3$.

Proof. It can be inferred from the Explanation of last example that

$$a_m = f(m) = \sum_{k=0}^{p} \Delta^k f(0) \binom{m}{k}.$$

Hence

$$\begin{aligned}\sum_{m=1}^{n} a_m &= \sum_{m=1}^{n} \left(\sum_{k=0}^{p} \Delta^k f(0) \binom{m}{k} \right) \\ &= \sum_{k=0}^{p} \Delta^k f(0) \sum_{m=1}^{n} \binom{m}{k}\end{aligned}$$

$$= \sum_{k=0}^{p} \Delta^k f(0)(C_k^k + \cdots + C_n^k)$$

$$= \sum_{k=0}^{p} \Delta^k f(0)(C_{k+1}^{k+1} + C_{k+1}^k + \cdots + C_n^k)$$

$$= \sum_{k=0}^{p} \Delta^k f(0)(C_{k+2}^{k+1} + C_{k+2}^k + \cdots + C_n^k)$$

$$= \cdots = \sum_{k=0}^{p} C_{n+1}^{k+1} \Delta^k f(0).$$

Therefore ① holds.

When $f(x) = x^3$,

$$\Delta f(x) = (x+1)^3 - x^3 = 3x^2 + 3x + 1,$$
$$\Delta^2 f(x) = 3(x+1)^2 + 3(x+1) + 1 - (3x^2 + 3x + 1) = 6x + 6,$$
$$\Delta^3 f(x) = 6(x+1) + 6 - (6x+6) = 6.$$

Hence $\Delta f(0) = 1$, $\Delta^2 f(0) = 6$, $\Delta^3 f(0) = 6$. Then by ①, we have

$$\sum_{m=1}^{n} m^3 = 6C_{n+1}^4 + 6C_{n+1}^3 + C_{n+1}^2 = \left(\frac{n(n+1)}{2}\right)^2.$$

Explanation. Here we have given the summing formula of the first n terms of a pth order arithmetic sequence (with the general formula provided). By this, we thus, give the summing formula of $\sum_{m=1}^{n} m^p \ (p = 1, 2, \cdots)$.

Example 4. Given the polynomial $f(x) = x^n + a_1 x^{n-1} + \cdots + a_n$, where $a_1, \cdots, a_n \in \mathbf{R}$. Prove that there is at least one number no less than $\frac{n!}{2^n}$ among

$$|f(1)|, |f(2)|, \cdots, |f(n+1)|.$$

Proof. It can be deduced from Theorem 1 that

$$\Delta^n f(x) = \sum_{i=0}^{n} (-1)^i C_n^i f(x + n - i). \qquad ①$$

Noting that $f(x) = x^n + a_1 x^{n-1} + \cdots + a_n$, we have $\Delta^n f(x) = n!$.

Letting $x = 1$ in ①, we have

$$n! = \sum_{i=0}^{n} (-1)^i C_n^i f(n+1-i). \quad ②$$

If the proposition is false, then for $0 \leqslant i \leqslant n$, there holds

$$|f(n+1-i)| < \frac{n!}{2^n}.$$

Along with ②, there must be

$$n! \leqslant \sum_{i=0}^{n} |(-1)^i C_n^i f(n+1-i)| < \sum_{i=0}^{n} C_n^i \cdot \frac{n!}{2^n} = n!,$$

which is contradictory.

Hence the proposition is true.

Explanation. This problem can also be dealt with Lagrange interpolation formula.

Example 5. For nonnegative integer N, let $u(N)$ be the number of 1s in the binary representation of N (for example, $u(10) = 2$, because $10 = (1010)_2$). Denote the degree of $p(x)$ by $\deg p(x)$. For any $k \in \mathbf{N}^*$, prove that it holds

$$\sum_{i=0}^{2^k-1} (-1)^{u(i)} p(i) = \begin{cases} 0, & \text{if } \deg p(x) < k; \\ (-1)^k \alpha \cdot k! \cdot 2^{\frac{k(k-1)}{2}}, & \text{if } \deg p(x) = k. \end{cases}$$

Here α is the leading coefficient of $p(x)$.

Proof. Solve the problem by the method of differences.

For $t \in \mathbf{N}^*$, let $\Delta_t(p(x)) = p(x) - p(x+t)$. Then

$$q_k(x) = \Delta_1(\Delta_2(\Delta_4 \cdots (\Delta_{2^{k-1}}(p(x)))\cdots))$$

is also a polynomial of x.

Let's prove the following by inducting on k.

$$\sum_{i=0}^{2^k-1} (-1)^{u(i)} p(i) = q_k(0). \quad ①$$

When $k = 1$, the left side of ① $= p(0) - p(1)$, while the right side

$q_1(0) = p(0) - p(1)$. Hence ① holds for $k = 1$.

Now suppose that ① holds for k. Let's consider the case of $k + 1$. Then

$$\begin{aligned}\sum_{i=0}^{2^{k+1}-1}(-1)^{u(i)}p(i) &= \sum_{i=0}^{2^k-1}(-1)^{u(i)}p(i) + \sum_{i=2^k}^{2^{k+1}-1}(-1)^{u(i)}p(i)\\ &= \sum_{i=0}^{2^k-1}(-1)^{u(i)}p(i) - \sum_{i=0}^{2^k-1}(-1)^{u(i)}p(2^k+i)\\ &= \sum_{i=0}^{2^k-1}(-1)^{u(i)}(p(i) - p(2^k+i))\\ &= \sum_{i=0}^{2^k-1}(-1)^{u(i)}\Delta_{2^k}(p(i)).\end{aligned}$$

Now substitute $\Delta_{2^k}(p(x))$ for $p(x)$ and apply the inductive hypothesis to it. It can be deduced that

$$\sum_{i=0}^{2^{k+1}-1}(-1)^{u(i)}p(i) = q_k(\Delta_{2^k}(p(0))) = q_{k+1}(0).$$

Therefore for $k \in \mathbf{N}^*$, ① holds.

Note that when $\deg(p(x)) \leqslant k$, every time we take the difference of $p(x)$, the degree is reduced by one. Hence when $\deg p(x) < k$, $q_k(x) = 0$. When $\deg p(x) = k$, for $t \in \mathbf{N}^*$, we have

$$\begin{aligned}\Delta_t(p(x)) &= p(x) - p(x+t)\\ &= \alpha(x^k - (x+t)^k) + \beta(x^{k-1} - (x+t)^{k-1}) + \cdots.\end{aligned}$$

By the Binomial Theorem, we can deduce that $\Delta_t(p(x))$ is a polynomial of degree $k - 1$ and the leading coefficient is $-\alpha tk$. Therefore $q_k(x)$ is a constant polynomial and

$$\begin{aligned}q_k(x) &= \left(\prod_{j=0}^{k-1}(-(j+1)\cdot 2^j)\right)\cdot\alpha\\ &= (-1)^k\cdot 2^{\frac{k(k-1)}{2}}\cdot k!\cdot\alpha.\end{aligned}$$

Hence the proposition is true.

Explanation. We can get a different identity when we take different polynomial $p(x)$ of degree k.

Example 6. Let $\{a_n\}$, $\{b_n\}$ be two sequences. Prove the following binomial inversion formula: for any $n \in \mathbf{N}^*$, $a_n = \sum_{k=0}^{n} C_n^k b_k$ holds if and only if for any $n \in \mathbf{N}^*$, $b_n = \sum_{k=0}^{n} (-1)^{n-k} C_n^k a_k$ holds.

Proof. For $m \in \mathbf{N}^*$, let $f(x)$ be a polynomial of degree m satisfying that for $0 \leqslant k \leqslant m$, $f(k) = a_k$.

From the difference polynomial in example 2, it can be deduced that

$$f(x) = \sum_{k=0}^{m} \Delta^k f(0) \binom{x}{k}.$$

Let $g(x) = \sum_{k=0}^{m} b_k \binom{x}{k}$. Then $g(x)$ is a polynomial of degree m.

If for any $n \in \mathbf{N}^*$, it holds that $a_n = \sum_{k=0}^{n} C_n^k b_k$. Then for any $0 \leqslant n \leqslant m$, it holds that $g(n) = \sum_{k=0}^{m} C_n^k b_k = a_n = f(n)$. This shows that there are $m+1$ different roots ($x = 0, 1, 2, \cdots, m$) of $f(x) - g(x)$. Hence it is a zero polynomial. Therefore $b_n = \Delta^n f(0)$. Along with Theorem 1, it can be deduced that

$$\begin{aligned} b_n = \Delta^n f(0) &= \sum_{k=0}^{n} (-1)^{n-k} C_n^k f(k) \\ &= \sum_{k=0}^{n} (-1)^{n-k} C_n^k a_k. \end{aligned}$$

Conversely, if for any $n \in \mathbf{N}^*$, it holds that $b_n = \sum_{k=0}^{n} (-1)^{n-k} C_n^k a_k$, then $b_n = \Delta^n f(0)$. Hence $g(x) = f(x)$ and thus,

$$a_n = f(n) = g(n) = \sum_{k=0}^{m} C_n^k b_k = \sum_{k=0}^{n} C_n^k b_k$$

(note that we have to take $m \geqslant n$).

In conclusion, the binomial inversion formula holds.

Explanation. We have given several formulas related to the method of differences. They are useful in proving some identities. They are also often applied in finding general formulas and summation of higher-order arithmetic sequences.

7 Recursive Sequences

If the nth term a_n of sequence $\{a_n\}$ is determined by several terms before it, then the sequence is a recursive sequence. In fact, arithmetic and geometric sequences are recursive sequences. The recurrence relation of them is $a_n = 2a_{n-1} - a_{n-2}$ and $a_n = a_{n-1} \cdot q$, respectively.

Generally, suppose

$$a_{n+k} = F(a_n, a_{n+1}, \cdots, a_{n+k-1}). \quad ①$$

That is, a_{n+k} is a function of a_n, a_{n+1}, $\cdots$, a_{n+k-1} and the initial data a_1, $\cdots$, a_k are determined. Then sequence $\{a_n\}$ is called a kth recursive sequence and ① is called the recurrence relation of $\{a_n\}$.

The problems related to recursive sequences fall into two classes. One is finding the general formula (or other properties) of a sequence, given the recurrence relation; the other is establishing the recurrence relation first and then seeking the essence of a problem by the idea of recurrence.

Now let's give some instrumental results.

Sequence $\{a_n\}$ satisfying the following recurrence relation is called a homogeneous linear recursive sequence with constant coefficients.

$$a_{n+k} = c_1 a_{n+k-1} + c_2 a_{n+k-2} + \cdots + c_k a_n, \quad ②$$

where c_1, c_2, $\cdots$, c_k are constant.

Note that if λ is a root of

$$\lambda^k = c_1 \lambda^{k-1} + c_2 \lambda^{k-2} + \cdots + c_k, \quad ③$$

then sequence $\{\lambda^n\}$ $(n = 1, 2, \cdots)$ satisfies recurrence relation ②. Moreover if the roots of ③ are distinct, letting them be λ_1, λ_2, $\cdots$, λ_k, then sequence $\{A_1\lambda_1^n + A_2\lambda_2^n + \cdots + A_k\lambda_k^n\}$ $(n = 1, 2, \cdots)$ satisfies ② and the coefficients A_1, A_2, $\cdots$, A_k can be determined by initial data a_1, a_2, $\cdots$, a_k (by solving a linear system). In this way, we get the general term of a sequence satisfying ② with given initial data a_1, a_2, $\cdots$, a_k.

This method of finding the general formula of a linear recursive sequence is called the method of characteristic roots. And ③ is called the characteristic equation of ②. The result will be rather complicated if there are multiple roots of ③, which will be illustrated in later examples.

In comparison, we can also make use of the method of generating functions. Generally, for sequence $\{a_n\}(n = 0, 1, 2, \cdots)$, the following formal series

$$f(x) = a_0 + a_1 x + a_2 x^2 + \cdots$$

is called the generating function of sequence $\{a_n\}$.

For example, the generating function of constant sequence $a_n = 1$, $n = 0, 1, 2, \cdots$ is $f(x) = 1 + x + x^2 + \cdots = \frac{1}{1-x}$ $(|x| < 1)$, which is the summing formula of infinite decaying geometric sequences.

Since the method of generating functions involves knowledge about advanced sequences like convergence of series, we will give the following formula of form series without proof and then show how to utilize this method in examples.

For $\alpha \in \mathbf{R}$, denote $\binom{\alpha}{n} = \frac{\alpha(\alpha-1)\cdots(\alpha-n+1)}{n!}$, $n \in \mathbf{N}$ (it is a generalization of binomial coefficients, which has been referred to in last section. Also, we define $\binom{\alpha}{0} = 1$). Then

$$(1+x)^\alpha = 1 + \binom{\alpha}{1}x + \binom{\alpha}{2}x^2 + \cdots. \qquad ④$$

Specifically, when $\alpha \in \mathbf{N}^*$, ④ is the binomial theorem.

There is no uniform method to deal with recursive sequences of other forms. One of the common methods is the method of fixed points.

Example 1. Given sequence $\{a_n\}$ satisfying that

$$a_1 = 0, \quad a_{n+1} = 5a_n + \sqrt{24a_n^2 + 1}, \quad n = 1, 2, \cdots,$$

find the general formula.

Proof. Transform the recurrence relation. It yields

$$(a_{n+1} - 5a_n)^2 = 24a_n^2 + 1.$$

Equivalently,

$$a_{n+1}^2 - 10a_n a_{n+1} + a_n^2 = 1. \quad ①$$

Substitute the subscript $n+1$ for n. It yields

$$a_{n+2}^2 - 10a_{n+1}a_{n+2} + a_{n+1}^2 = 1. \quad ②$$

Comparing ① with ②, we can deduce that a_n and a_{n+2} are the roots of equation

$$x^2 - 10a_{n+1}x + a_{n+1}^2 - 1 = 0. \quad ③$$

It can be deduced from the recurrence relation that $\{a_n\}$ is an increasing sequence. Hence a_n and a_{n+2} are distinct. Therefore applying Vieta's theorem to ③, we get

$$a_{n+2} + a_n = 10a_{n+1},$$

equivalently $a_{n+2} = 10a_{n+1} - a_n$, $n = 1, 2, \cdots$.

This is essentially a problem of second order homogeneous linear recursive sequence. We will give the following two ways to find the general formulas.

Method 1. The characteristic equation is

$$\lambda^2 = 10\lambda - 1,$$

with two roots $\lambda_{1,2} = 5 \pm 2\sqrt{6}$. Hence we can assume

$$a_n = A \cdot (5 + 2\sqrt{6})^n + B \cdot (5 - 2\sqrt{6})^n.$$

Noting the initial data $a_1 = 0$, we have $a_2 = 1$. Then solving

$$\begin{cases}(5+2\sqrt{6})A+(5-2\sqrt{6})B=0,\\(5+2\sqrt{6})^2A+(5-2\sqrt{6})^2B=1,\end{cases}$$

we get $A=\dfrac{5-2\sqrt{6}}{4\sqrt{6}}$, $B=\dfrac{-5-2\sqrt{6}}{4\sqrt{6}}$. Therefore

$$a_n=\frac{1}{4\sqrt{6}}((5+2\sqrt{6})^{n-1}-(5-2\sqrt{6})^{n-1}).$$

Method 2. Utilize the method of generating functions. For the sake of convenience, we can define complementally that $a_0=-1$ by the recurrence relation and the condition that $a_1=0$, $a_2=1$. Then the generating function of $\{a_n\}$ $(n=0,1,2,\cdots)$ satisfies that

$$\begin{aligned}f(x)&=\sum_{n=0}^{+\infty}a_nx^n=-1+\sum_{n=2}^{+\infty}a_nx^n\\&=-1+10\sum_{n=2}^{+\infty}a_{n-1}x^n-\sum_{n=2}^{+\infty}a_{n-2}x^n\\&=-1+10x\sum_{n=1}^{+\infty}a_nx^n-x^2\sum_{n=0}^{+\infty}a_nx^n\\&=-1+10x(f(x)+1)-x^2f(x).\end{aligned}$$

Solving the equation, we get $f(x)=\dfrac{10x-1}{x^2-10x+1}$.

Letting $f(x)=\dfrac{A}{1-(5+2\sqrt{6})x}+\dfrac{B}{1-(5-2\sqrt{6})x}$ (rewrite $f(x)$ in the form of a partial fraction), we have

$$\begin{cases}(5-2\sqrt{6})A+(5+2\sqrt{6})B=-10,\\A+B=-1.\end{cases}$$

Then $B=\dfrac{1}{4\sqrt{6}}(-5-2\sqrt{6})$, $A=\dfrac{1}{4\sqrt{6}}(5-2\sqrt{6})$.

Now expand $f(x)$ in the form of formal series.

$$f(x)=A\sum_{n=0}^{+\infty}(5+2\sqrt{6})^nx^n+B\sum_{n=0}^{+\infty}(5-2\sqrt{6})^nx^n$$

$$= \sum_{n=0}^{+\infty} (A(5+2\sqrt{6})^n + B(5-2\sqrt{6})^n)x^n.$$

Hence

$$\begin{aligned} a_n &= A(5+2\sqrt{6})^n + B(5-2\sqrt{6})^n \\ &= \frac{1}{4\sqrt{6}}((5+2\sqrt{6})^{n-1} - (5-2\sqrt{6})^{n-1}). \end{aligned}$$

Explanation. This example demonstrates fundamental steps of finding the general formula of sequences by the method of generating functions: first, find the generating function $f(x)$ by the recurrence relation, then expand $f(x)$ in the form of formal series and get the general formula from the equality of the coefficients of corresponding terms.

Example 2. Sequence $\{a_n\}$ satisfies that $a_1 = 2$ and that for $n = 1, 2, \cdots$,

$$a_{n+1} = \frac{a_n}{2} + \frac{1}{a_n}. \quad ①$$

Find the general formula of the sequence.

Solution. Here we introduce the method of fixed points (which comes from the idea of iteration of functions). First, find the roots of equation

$$\lambda = \frac{\lambda}{2} + \frac{1}{\lambda}. \quad ②$$

They are $\lambda_{1,2} = \pm\sqrt{2}$.

Noting ① and that $a_1 = 2$, we can deduce that each term of $\{a_n\}$ is a positive rational number. Now by ① − ②, we have that for $\lambda = \pm\sqrt{2}$, it holds that

$$a_{n+1} - \lambda = \frac{a_n - \lambda}{2} + \left(\frac{1}{a_n} - \frac{1}{\lambda}\right),$$

which can be transformed into

$$\frac{a_{n+1}-\lambda}{a_n-\lambda}=\frac{1}{2}-\frac{1}{\lambda a_n}=\frac{\lambda a_n-2}{2\lambda a_n}. \quad ③$$

Take $\lambda=\sqrt{2}$, $-\sqrt{2}$ in ③ respectively. The quotient of both sides of the two formulas is

$$\frac{a_{n+1}-\sqrt{2}}{a_{n+1}+\sqrt{2}}=\left(\frac{a_n-\sqrt{2}}{a_n+\sqrt{2}}\right)^2.$$

Hence we have

$$\begin{aligned}\frac{a_{n+1}-\sqrt{2}}{a_{n+1}+\sqrt{2}}&=\left(\frac{a_n-\sqrt{2}}{a_n+\sqrt{2}}\right)^2=\left(\frac{a_{n-1}-\sqrt{2}}{a_{n-1}+\sqrt{2}}\right)^{2^2}\\&=\cdots=\left(\frac{a_1-\sqrt{2}}{a_1+\sqrt{2}}\right)^{2^n}=(3-2\sqrt{2})^{2^n}\\&=(\sqrt{2}-1)^{2^{n+1}}.\end{aligned}$$

Therefore $\dfrac{a_n-\sqrt{2}}{a_n+\sqrt{2}}=(\sqrt{2}-1)^{2^n}$. Solving the equation, we get

$$a_n=\frac{\sqrt{2}\left(1+(\sqrt{2}-1)^{2^n}\right)}{1-(\sqrt{2}-1)^{2^n}}.$$

Example 3. Let m, $n\in\mathbf{N}^*$ and mn be a triangle number (i.e., there exists $t\in\mathbf{N}^*$, satisfying that $mn=1+2+\cdots+t$). Prove that there exists a positive integer k such that for any subscript j, the sequence $\{a_n\}$ defined by the following recurrence relation

$$a_1=m,\ a_2=n,\ a_j=6a_{j-1}-a_{j-2}+k,\ j=3,\ 4,\ \cdots$$

satisfies that a_ja_{j+1} is a triangle number.

Proof. Note that

x is a triangle number $\Leftrightarrow$ there exists $t\in\mathbf{N}^*$, satisfying that $x=1+2+\cdots+t$

$\Leftrightarrow$there exists $t\in\mathbf{N}^*$ satisfying that $x=\dfrac{t(t+1)}{2}$

$\Leftrightarrow$there exists $t\in\mathbf{N}^*$ satisfying that $8x+1=(2t+1)^2$.

Hence we need only prove that there exists $k \in \mathbf{N}^*$, satisfying that for any $j \in \mathbf{N}^*$, $8a_j a_{j+1} + 1$ is a perfect square.

Starting from the formula of the perfect square, let's see whether there exists $k \in \mathbf{N}^*$, satisfying that for any $j \in \mathbf{N}^*$, it holds that $8a_j a_{j+1} + 1 = (a_j + a_{j+1} + l)^2$, where l is a constant related to k only.

Taking $j = 1$ in the conjecture, we have that $l = \sqrt{8mn+1} - m - n$ (since mn is a triangle number, $\sqrt{8mn+1} \in \mathbf{N}^*$).

Moreover, if for any $j \in \mathbf{N}^*$, it holds that

$$8a_j a_{j+1} + 1 = (a_j + a_{j+1} + l)^2. \qquad ①$$

Then substitute $j+1$ for j in ①. It yields

$$8a_{j+1} a_{j+2} + 1 = (a_{j+1} + a_{j+2} + l)^2. \qquad ②$$

Subtracting both sides of the two formulas, we get

$$\begin{aligned} &8(a_{j+2} - a_j)a_{j+1} = (a_{j+2} - a_j)(a_{j+2} + 2a_{j+1} + a_j + 2l) \\ \Leftrightarrow\ &8a_{j+1} = a_{j+2} + 2a_{j+1} + a_j + 2l \qquad ③ \\ \Leftrightarrow\ &a_{j+2} = 6a_{j+1} - a_j - 2l. \end{aligned}$$

Then it can be deduced from ①, ③ that ② holds.

With the analysis above, if we let

$$k = -2l = 2(m+n) - 2\sqrt{8mn+1},$$

then the sequence $\{a_n\}$ defined by the given recurrence formula satisfies the condition, which can be proved by mathematical induction.

In conclusion, the proposition is true.

Explanation. Here we applied the idea of proving after guessing, which is not unique to problems of sequences, but common throughout the study of math. It is a reflection of inspiration.

Example 4. Sequence 0, 1, 3, 0, 4, 9, 3, 10, ⋯, is defined as follows:

$a_0 = 0$ and for $n = 1, 2, \cdots$, there holds

$$a_n = \begin{cases} a_{n-1} - n, & \text{if } a_{n-1} \geqslant n, \\ a_{n-1} + n, & \text{rest.} \end{cases}$$

Does every nonnegative integer occur in this sequence? Prove your result.

Solution. Every nonnegative integer occurs in this sequence.

Noting that by definition, $\{a_n\}$ is a sequence of integers, let's first determine the range of each term in $\{a_n\}$. Inducting on n, let's prove that

$$0 \leqslant a_n \leqslant 2n - 1 \ (n \geqslant 1). \quad ①$$

When $n = 1$, it can be inferred from the condition that $a_1 = 1$. Then ① holds. Suppose that a_{n-1} $(n \geqslant 2)$ satisfies ①. When $n \leqslant a_{n-1} \leqslant 2n - 3$, $a_n = a_{n-1} - n \in [0, n-3]$ (noting that $n \leqslant 2n-3$, we have $n \geqslant 3$), which satisfies ①; when $0 \leqslant a_{n-1} \leqslant n-1$, $a_n = a_{n-1} + n \in [n, 2n-1]$, which also satisfies ①. Therefore ① holds for $n \in \mathbf{N}^*$.

Now let's rewrite the recurrence formula of a_n: when $a_{n-1} = 0$, $a_n = n$, $a_{n+1} = 2n+1$; when $a_{n-1} \in [1, n-1]$, $a_n = n + a_{n-1} \in [n+1, 2n-1]$, and then $a_{n+1} = a_n - (n+1) = a_{n-1} - 1$; when $a_{n-1} \in [n, 2n-3]$, $a_n = a_{n-1} - n \in [0, n-3]$, and then $a_{n+1} = a_n + (n+1) = a_{n-1} + 1$. Hence when $n \geqslant 3$, we have that

$$a_{n+1} = \begin{cases} 2n+1, & \text{if } a_{n-1} = 0, \\ a_{n-1} - 1, & \text{if } a_{n-1} \in [1, n-1], \\ a_{n-1} + 1, & \text{if } a_{n-1} \in [n, 2n-3]. \end{cases} \quad ②$$

As to the original question, if there are nonnegative numbers that don't occur in the sequence, we can take the least one. Let it be M, then $M > 1$ and $M - 1$ occurs in the sequence. Suppose that $a_{n-1} = M - 1$. If $a_{n-1} \in [n, 2n-3]$, then $M = a_{n-1} + 1 = a_{n+1}$, which is contradictory. Hence $a_{n-1} \leqslant n-1$. Noting that $M > 1$, we have $a_{n-1} \in [1, n-1]$ and then $a_{n+1} = a_{n-1} - 1 \in [0, n-2]$. Moreover, $a_{n+3} = a_{n+1} - 1$ (or $a_{n+1} = 0$). Repeating the steps again and again, we get a subsequence $a_{n-1} > a_{n+1} > a_{n+3} > \cdots > a_{s-1} = 0$, where $s \geqslant n+2$.

Noting that $a_{n-1} \leqslant n-1$, we have $M \leqslant n$. Since $a_{s-1} = 0$, by ②, we can deduce that $a_{s+1} = 2s + 1 > s + 2$. Moreover, it holds that

$$a_{s+2} = a_{s+1} - (s+2) = s - 1 \in [0, s+1].$$

We can infer in the same way that $a_{s+1} \in \{0, a_{s+2}-1\}$, $\cdots$. Therefore there must be a subscript t such that $a_t = M$ (since $s-1 \geqslant n+1 \geqslant M$). Hence M must be a term in $\{a_n\}$, which is contradictory.

In conclusion, every nonnegative integer appears in $\{a_n\}$.

Explanation. The key to this question is to rewrite the recurrence formula given in a proper way into the form of ②, which reveals the feature of adding 1 or subtracting 1 every two terms. This laid a solid foundation for proving that every nonnegative integer, occurs in the sequence.

Example 5. Let A_n denote the set of n-letter words consists of a, b, c with no consecutive as and bs; let B_n denote the set of n-letter words consists of a, b, c with no three distinct consecutive letters. For any positive integer n, prove that $|B_{n+1}| = 3|A_n|$ holds.

Proof. Let's apply the method of recurrence to solve this question.

Let c_n denote the number of words with initial c and d_n denote the number of words with initial a or b in A_n.

For words in A_{n+1}, we can classify them by their initials. If the initial is c, then the word belongs to A_n as we remove the first letter; if the initial is a, then the second letter must be c or b; if the initial is b, then the second letter must be c or a. Therefore the following recurrence formula holds:

$$\begin{cases} c_{n+1} = |A_n| = c_n + d_n, \\ d_{n+1} = 2c_n + d_n. \end{cases} \quad ①$$

Now let c'_n denote the number of the words in B_n whose first two letters are same and d'_n denote the number of the words in B_n whose first two letters are different.

For words in B_{n+1}, we classify them by the first two letters. If they are same, then we can give the third letter arbitrarily. The word belongs to B_n as we remove the first letter. If they are different, then the third letter must be same to one of the first two letters. If it is same to the first letter, we can get d'_n words as we remove the first letter. If it is same to the second letter, we can get $2c'_n$ words, as we remove the first letter (here the coefficient is 2 because we get the same word as we remove the first letter from $abb\cdots$ and $cbb\cdots$). Therefore the recurrence formula is

$$\begin{cases} c'_{n+1} = |B_n| = c'_n + d'_n, \\ d'_{n+1} = 2c'_n + d'_n. \end{cases} \qquad ②$$

Noting that the recurrence formulas ① and ② are exactly the same. The only difference is there initial value. By enumerating directly, we have $c_1 = 1$, $d_1 = 2$; $c'_2 = 3$, $d'_2 = 6$. Hence $c'_2 = 3c_1$, $d'_2 = 3d_1$. From the recurrence formula, we can deduce that $c'_{n+1} = 3c_n$, $d'_{n+1} = 3d_n$. Noting that $|A_n| = c_{n+1}$ and $|B_n| = c'_{n+1}$, we get $|B_{n+1}| = 3|A_n|$.

The proposition is proved.

Explanation. It is an important method to apply the idea of recurrence to combination and counting problems. The recurrence formula we established here can be turned into homogeneous linear recurrence relation with constant coefficients. We can then find the value of $|A_n|$.

Example 6. Sequence of real numbers $\{a_n\}$ satisfies that for any different positive integers i, j, $|a_i - a_j| \geqslant \dfrac{1}{i+j}$ holds. Moreover, there exists a real number c such that for any $n \in \mathbf{N}^*$, $0 \leqslant a_n \leqslant c$ holds.

Prove that $c \leqslant 1$.

Proof. This question does not give the relationship between the terms of the sequence in the form of an equation. Instead, it describes the distance of the terms by an inequality. There is a sense of real analysis in the process of solving the question. The solution is based on the idea of summation after splitting terms.

For $n \geqslant 2$, let $\pi(1)$, $\cdots$, $\pi(n)$ be a permutation of $1, 2, \cdots, n$, such that

$$0 \leqslant a_{\pi(1)} < a_{\pi(2)} < \cdots < a_{\pi(n)} \leqslant c. \qquad ①$$

Note that the terms of $\{a_n\}$ are distinct according to the conditions. Then $a_1, \cdots, a_n$ are just sorted from the least to the greatest in ①.

From ① and the conditions, we can deduce that

$$c \geqslant a_{\pi(n)} - a_{\pi(1)} = \sum_{k=1}^{n-1}(a_{\pi(k+1)} - a_{\pi(k)})$$
$$\geqslant \sum_{k=1}^{n-1} \frac{1}{\pi(k+1)+\pi(k)}.$$

By Cauchy inequality, we have

$$\sum_{k=1}^{n-1} \frac{1}{\pi(k+1)+\pi(k)} \geqslant \frac{(n-1)^2}{\sum_{k=1}^{n-1}(\pi(k+1)+\pi(k))}$$
$$= \frac{(n-1)^2}{2\sum_{k=1}^{n}\pi(k) - \pi(1) - \pi(n)} = \frac{(n-1)^2}{n(n+1) - \pi(1) - \pi(n)}$$
$$\geqslant \frac{(n-1)^2}{n(n+1)-1-2} > \frac{(n-1)^2}{n(n+1)-2} = \frac{n-1}{n+2} = 1 - \frac{3}{n+2}.$$

Hence we have

$$c \geqslant 1 - \frac{3}{n+2}.$$

Letting $n \to +\infty$, we get $c \geqslant 1$.

The proposition is proved.

Example 7. An infinite sequence of real numbers $\{a_n\}$ is defined as follows: a_0, a_1 are two different positive real numbers and $a_n = |a_{n+1} - a_{n+2}|$, $n = 0, 1, 2, \cdots$. Is it possible that the sequence is bounded? Please prove your result.

Solution. This sequence must be unbounded. The fundamental idea is to take an increasing unbounded subsequence from $\{a_n\}$.

In fact, if there exists $n \in \mathbf{N}^*$, satisfying that $a_n = a_{n+1}$, then by the recurrence formula, we have $a_{n-1} = 0$, and moreover $a_{n-2} = a_{n-3}$ (note that each term of $\{a_n\}$ is nonnegative). As we deduce successively, we have $a_1 = a_2$ or one of a_1, a_2 is zero, which is contrary to the condition that a_1, a_2 are two different real numbers. Therefore for any $n \in \mathbf{N}^*$, $a_n \neq a_{n+1}$ (equivalently, there are no same consecutive terms in $\{a_n\}$). Hence noting the recurrence formula, for any $n \in \mathbf{N}^*$, $a_n > 0$.

Now let's find an increasing subsequence $\{b_m\}$ from $\{a_n\}$.

It can be inferred from the condition that $a_{n+2} = a_n + a_{n+1}$ or $a_{n+2} = a_{n+1} - a_n$. If the former holds, then $a_{n+2} > a_{n+1}$. If the latter holds, then $a_{n+2} < a_{n+1}$. It must hold $a_{n+3} = a_{n+2} + a_{n+1}$, since $a_{n+2} = a_{n+1} - a_n$ (otherwise $a_{n+3} = a_{n+2} - a_{n+1} < 0$, which is contradictory). Therefore $a_{n+3} > a_{n+1}$. The discussion shows that it holds either $a_{n+2} > a_{n+1}$, or $a_{n+2} < a_{n+1} < a_{n+3}$.

With the result above, we can remove from $\{a_n\}$ all the terms a_{n+1} satisfying that $a_{n+1} < a_n$ and $a_{n+1} < a_{n+2}$ (note that when $n \geqslant 2$, $a_{n+2} > a_n$). It's certain that we remove a_1 and keep a_2 if $a_1 > a_2$. As we remove the terms, the remaining ones can be denoted by b_1, b_2, $\cdots$ and then sequence $\{b_m\}$ is increasing.

At last, let's prove that $\{b_m\}$ is unbounded.

For any $m \in \mathbf{N}^*$, it suffices to prove that $b_{m+2} - b_{m+1} \geqslant b_{m+1} - b_m$ (since by summing after splitting the terms, we have $b_{m+2} - b_2 \geqslant m(b_2 - b_1)$. Letting $m \to +\infty$, we get that $\{b_m\}$ is unbounded).

By the definition of $\{b_m\}$, we can set $b_{m+2} = a_{n+2}$ (note that n may not be same to m). Since a_{n+2} is not removed, $a_{n+2} > a_{n+1}$. If $a_{n+1} > a_n$, then $b_{m+1} = a_{n+1}$ while $b_m = a_n$ or a_{n-1} (if the former, then $a_n > a_{n-1}$). Thus, it always holds that $b_m \geqslant a_{n-1}$. Therefore we have

$$b_{m+2} - b_{m+1} = a_{n+2} - a_{n+1} = a_n = a_{n+1} - a_{n-1} \geqslant b_{m+1} - b_m.$$

If $a_{n+1} < a_n$, then $b_{m+1} = a_n$ while $b_m = a_{n-1}$ or a_{n-2} (if the latter, then $a_{n-2} > a_{n-1}$. Otherwise a_{n-1} is not removable). Hence

$$b_{m+2} - b_{m+1} = a_{n+2} - a_n = a_{n+1} = a_n - a_{n-1} \geqslant b_{m+1} - b_m.$$

The proposition is proved.

Explanation. Readers are encouraged to write a specific sequence when reading the answer which helps to find the relationship between $\{a_n\}$ and $\{b_m\}$. Similar to the previous question, this recursive sequence is not defined by definite formulas. Both questions involve the estimate of inequality, which is a reflection of analysis.

Example 8. Sequence $\{a_n\}$ satisfies the recurrence formula

$$a_{n+1} = \frac{a_n^2 - 1}{n+1}, \ n = 0, 1, 2, \cdots.$$

Is there a positive real number such that both of the following results hold?

(1) If $a_0 \geqslant a$, then $\lim\limits_{n\to\infty} a_n$ does not exist;

(2) If $0 < a_0 < a$, then $\lim\limits_{n\to\infty} a_n = 0$.

Solution. There exists such positive real number $a = 2$. Mathematical induction is applied repeatedly in the solution to this question. The details are left to readers.

(1) When $a_0 \geqslant 2$, we can prove by mathematical induction that for $n \geqslant 0$, $a_n \geqslant n + 2$ holds. Then $\lim\limits_{n\to\infty} a_n$ does not exist.

(2) when $0 < a_0 < 2$, it can be divided into two cases:

Case 1. $0 < a_0 \leqslant 1$. Now we can prove by mathematical induction that for any $n \in \mathbf{N}^*$, $|a_n| \leqslant \frac{1}{n}$ holds; thus, $\lim\limits_{n\to\infty} a_n = 0$.

Case 2. $1 < a_0 < 2$. If there exists $m \in \mathbf{N}^*$ satisfying that $a_{m+1} \leqslant 0$, we can take the least m. Then $0 < a_m \leqslant 1$ and $|a_{m+1}| = \frac{1 - a_m^2}{m+1} \leqslant \frac{1}{m+1}$. With this result, by mathematical induction, when $n \geqslant m + 1$, we can prove that $|a_n| \leqslant \frac{1}{n}$ holds. Hence $\lim\limits_{n\to+\infty} a_n = 0$.

Lastly, if for any $m \in \mathbf{N}^*$, $a_m > 0$ holds, noting that $1 < a_0 < 2$, we have $a_n > 1$ holds for $n \geqslant 0$. Now let $a_0 = 2 - \varepsilon$ $(0 < \varepsilon < 1)$. By the recurrence formula and mathematical induction, for any $n \in \mathbf{N}^*$, we can prove that $a_n < n + 2 - n\varepsilon$. Therefore taking $m = \left\lceil \frac{1}{\varepsilon} \right\rceil$, we have $a_m <$

$m+2-m\varepsilon \leqslant m+1$. Then we can prove by mathematical induction that for any $n > m$, $a_n \leqslant \frac{(m+1)^2-1}{n-1}$. When n is sufficiently large, it holds $a_n \leqslant 1$, which is contradictory. Hence there must be an $n \in \mathbf{N}^*$ satisfying that $a_n \leqslant 0$, which falls in the previous case.

In conclusion, $a = 2$ satisfies the conditions.

8 Periodic Sequences

We call sequence $\{a_n\}$ a periodic sequence, if there exist positive integers T and n_0, such that $a_n = a_{n+T}$ holds for any $n \geqslant n_0$. Moreover, we call $\{a_n\}$ a pure periodic sequence if $n_0 = 1$ and T the period of $\{a_n\}$.

By the definition of periodic sequence, if T is a period of $\{a_n\}$, then mT is also a period of $\{a_n\}$ for any $m \in \mathbf{N}^*$. Combining the property above with the Bezout's Theorem, a famous theorem in number theory, we can get the following theorem:

Theorem 1. If T_1 and T_2 are periods of periodic sequence $\{a_n\}$, then (T_1, T_2) (the greatest common divisor of T_1 and T_2) is also a period of $\{a_n\}$.

From this theorem, we can infer that a periodic sequence $\{a_n\}$ has its least positive period, which is in stark contrast with the fact that a periodic function f may not have a least positive period.

For sequence of integers $\{a_n\}$, it can be a periodic sequence modulo m, for some positive integer m, while it may not be a periodic sequence itself. This is the concept of modulo periodic sequence. Then there exist T, $n_0 \in \mathbf{N}^*$, such that for any $n \geqslant n_0$, $a_{n+T} \equiv a_n \pmod{m}$ holds.

Theorem 2. If sequence of integers $\{a_n\}$ is a recursive sequence with constant coefficients, then for any $m \in \mathbf{N}^*$, $\{a_n\}$ is always a periodic sequence modulo m.

Actually, if $\{a_n\}$ is a recursive sequence of degree k with constant coefficients, let's consider the following arrays

$$(a_1,\ a_2,\ \cdots,\ a_k),\ (a_2,\ a_3,\ \cdots,\ a_{k+1}),\ \cdots. \qquad ①$$

Each x_i in the array $(a_1, a_2, \cdots, a_k)$, can only take a value from $0, 1, 2, \cdots, m-1$, modulo m, then there are at most m^k different cases for the arrays in ①. Consequently, there exist $r, t \in \mathbf{N}^*$ (r < t), such that

$$(a_r,\ a_{r+1},\ \cdots,\ a_{r+k}) \equiv (a_t,\ a_{t+1},\ \cdots,\ a_{t+k}) \pmod m.$$

Let $T = t - r$. Noting that $\{a_n\}$ is a recursive sequence with constant coefficients, we can get that for any $n \geqslant r$, $a_{n+T} \equiv a_n \pmod m$.

Therefore, Theorem 2 holds.

Example 1. Let x_0, x_1 be positive real numbers and sequence $\{x_n\}$ satisfying that $x_{n+2} = \dfrac{4\max\{x_{n+1},\ 4\}}{x_n}$, $n = 0, 1, 2, \cdots$. Find the value of x_{2011}.

Solution. For the sake of convenience, let $x_n = 4y_n$. Then

$$y_{n+2} = \frac{\max\{y_{n+1},\ 1\}}{y_n},\ n = 0,\ 1,\ 2,\ \cdots.$$

We can get the following table by direct calculation:

	$y_0 \leqslant 1,\ y_1 \leqslant 1$	$y_0 \leqslant 1,\ y_1 > 1$	$y_0 > 1,\ y_1 \leqslant 1$	$y_0 > 1,\ y_1 > 1$
$y_2 =$	$\dfrac{1}{y_0}$	$\dfrac{y_1}{y_0}$	$\dfrac{1}{y_0}$	$\dfrac{y_1}{y_0}$
$y_3 =$	$\dfrac{1}{y_0 y_1}$	$\dfrac{1}{y_0}$	$\dfrac{1}{y_1}$	$\max\left\{\dfrac{1}{y_0},\ \dfrac{1}{y_1}\right\}$
$y_4 =$	$\dfrac{1}{y_1}$	$\dfrac{1}{y_1}$	$\dfrac{y_0}{y_1}$	$\dfrac{y_0}{y_1}$
$y_5 =$	y_0	y_0	y_0	y_0
$y_6 =$	y_1	y_1	y_1	y_1

Therefore $\{y_n\}$ is a pure periodic sequence with period of 5, so is $\{x_n\}$. As a result, $x_{2011} = x_1$.

Explanation. The recurrence relation given in the question above is a special form of Lyness Equation. Here determining a period by direct calculation is a straightforward and effective method to deal with fractional (periodic) recursive sequences.

Example 2. Let $0 \leqslant x_0 < 1$ and sequence $\{x_n\}$ satisfy

$$x_{n+1} = \begin{cases} 2x_n - 1, \text{if } \frac{1}{2} \leqslant x_n < 1, \\ 2x_n, \quad \text{if } 0 \leqslant x_n < \frac{1}{2}. \end{cases} \quad (n = 0, 1, 2, \cdots)$$

And $x_5 = x_0$. How many sequences are there that satisfy these conditions?

Solution. Note that sequence $\{x_n\}$ is determined uniquely when x_0 is fixed. Then we have converted the problem into finding how many different values that x_0 can take on.

We use the binary system to solve this question. Represent $\{x_n\}$ in binary numbers. Let $x_n = (0.b_1b_2\cdots)_2$. If $b_1 = 1$, then $\frac{1}{2} \leqslant x_n < 1$, and thus, $x_{n+1} = 2x_n - 1 = (0.b_2b_3\cdots)_2$; if $b_1 = 0$, then $0 \leqslant x_n < \frac{1}{2}$ and thus, $x_{n+1} = 2x_n = (0.b_2b_3\cdots)_2$. It indicates that $x_{n+1} = (0.b_2b_3\cdots)_2$ holds as long as $x_n = (0.b_1b_2\cdots)_2$ (which is equivalent to "swallowing" the first number after the decimal point).

Now, let $x_0 = (0.a_1a_2\cdots)_2$.

Then we can deduce that $x_5 = (0.a_6\,a_7\cdots)_2$ from our discussion above. Noting that $x_5 = x_0$, we have that x_0 is a recurring decimal in binary numbers, i. e., $x_0 = (0.\dot{a}_1a_2\cdots\dot{a}_5)_2 = \frac{(a_1\cdots a_5)_2}{2^5 - 1}$, in which $(a_1\cdots a_5)_2$ represents a nonnegative integer in binary numbers (noting that $a_1, \cdots, a_5$ are not all 1).

In conclusion, there are $2^5 - 1 = 31$ different values for x_0 (because $a_1, \cdots, a_5$ can take on 0 or 1 as their value, however, they are not all 1). That's to say, there are 31 different sequences.

Explanation. Here we use binary representation to turn recurrence relations into formulas with more regularity. Then combining it with the periodicity of the sequence, we grasp the structure of this sequence. In essence, we are mapping functions between different spaces.

Example 3. Let $f(x)$ be a polynomial with integer coefficients, and sequence $\{a_n\}$ is defined as follows.

$$a_0 = 0,\ a_{n+1} = f(a_n),\ n = 0,\ 1,\ 2,\ \cdots.$$

If $\{a_n\}$ is a pure periodic sequence, prove that the least positive period of $\{a_n\}$ is no more than 2.

Proof. We can turn this question into proving that if there exist $m \in \mathbf{N}^*$ such that $a_m = 0$, then $a_1 = 0$ or $a_2 = 0$.

According to the factor theorem, since $f(x)$ is a polynomial with integer coefficients, for any m, $n \in \mathbf{Z}(m \neq n)$, $m - n \mid f(m) - f(n)$ holds.

Now let $b_n = a_{n+1} - a_n$, $n = 0, 1, 2, \cdots$, then by the result above and the definition of $\{a_n\}$, we have $b_n \mid b_{n+1}$ (Note that if $b_n = 0$, then $b_{n+1} = f(a_{n+1}) - f(a_n) = 0$).

Since $a_m = a_0 = 0$, $a_{m+1} = f(a_0) = a_1$. Consequently, $b_m = b_0$.

If $b_0 = 0$, then $a_0 = a_1 = \cdots = a_m$ and the proposition is true. Otherwise, $|b_0| = |b_m| \neq 0$, then noting that $b_0 \mid b_1$, $b_1 \mid b_2$, $\cdots$, $b_{m-1} \mid b_m$, we can get $|b_0| = |b_1| = \cdots = |b_m|$.

Since

$$b_0 + b_1 + \cdots + b_{m-1} = a_m - a_0 = 0,$$

half of b_0, b_1, $\cdots$, b_{m-1} are positive integers, while the others negative integers. Consequently, there exists $k \in \{1, 2, \cdots, m-2\}$ such that $b_{k-1} = -b_k$, and then $a_{k-1} = a_{k+1}$. By the definition of $\{a_n\}$, we have $a_{n+2} = a_n$ holds for all $n \geqslant k-1$. Let $n = m$, we get

$$a_0 = a_m = a_{m+2} = f(a_{m+1}) = f(f(a_m)) = f(f(a_0)) = a_2.$$

Then $a_2 = 0$.

Hence the proposition is true.

Example 4. Let m be a positive integer greater than 1, sequence

$\{x_n\}$ is defined as follows: $x_1 = 1$, $x_2 = 2$, $\cdots$, $x_m = m$, and

$$x_{n+m} = x_{n+m-1} + x_n, \; n = 1, 2, \cdots. \qquad ①$$

Prove that there exist $m - 1$ consecutive terms in $\{x_n\}$ such that they are all multiples of m.

Proof. Consider sequence $\{x_k \pmod m\}$, in which $x_k \pmod m$, denoted by y_k, represents the remainder that x_k is divided by m. We turn to prove that there are $m - 1$ consecutive zeros in the sequence $\{y_n\}$.

By Theorem 2 and ①, we have that there exist n_0 and $T \in \mathbf{N}^*$ such that $y_{k+T} = y_k$ for any $k \geqslant n_0$. Specifically, we have

$$y_{n_0+m-1} = y_{n_0+m-1+T}, \; y_{n_0+m-2} = y_{n_0+m-2+T}.$$

Subtracting the two formulas and noting ① and the definition of y_k, we can get $y_{n_0-1} = y_{n_0-1+T}$, and $y_k = y_{k+T}$ for every $k \geqslant 1$ in the same manner.

To get our result and calculate conveniently, we extend the subscript of $\{x_n\}$ to negative integers according to the recurrence relation determined by ①. Combining this with what we discussed above, we can get that $y_k = y_{k+T}$, for every $k \in \mathbf{Z}$.

Now, by $x_n = x_{n+m} - x_{n+m-1}$, we can conclude that $x_0 = x_{-1} = \cdots = x_{-(m-2)} = 1$ (here we used the initial condition that $x_j = j$, for any $1 \leqslant j \leqslant m$), furthermore, $x_{-(m-1)} = x_{-m} = \cdots = x_{-(2m-3)} = 0$. Noting that $y_k = y_{k+T}$, we can get

$$(y_{-(2m-3)+T}, \cdots, y_{-(m-1)+T}) = (y_{-(2m-3)}, \cdots, y_{-(m-1)}) = (0, \cdots, 0).$$

Nevertheless, $y_{-(m-2)} = \cdots = y_0 = 1$. Therefore $-(2m-3) + T \geqslant 1$, which shows that there are $m - 1$ consecutive zeros among the terms with positive subscripts in sequence $\{y_n\}$.

Hence the proposition is true.

Example 5. Let m be a given positive integer, and for any positive integer n, $S_m(n)$ denote the sum of mth power of every digit of n in the decimal system. For instance, $S_3(172) = 1^3 + 7^3 + 2^3 = 352$.

Consider this sequence: n_0 is a positive integer, $n_k = S_m(n_{k-1})$, $k = 1, 2, \cdots$.

(1) For any positive integer n_0, prove that sequence $\{n_k\}$ is always a periodic sequence;

(2) When n_0 varies, prove that the set made up of the least positive periods in (1) is a finite set.

Proof. Note that for positive integer $n \geqslant 10^{m+1}$, there exists $p \in \mathbf{N}^*$, $p \geqslant m+1$ such that $10^p \leqslant n < 10^{p+1}$. Then n is a number with $p+1$ digits in the decimal system. Therefore

$$\begin{aligned} S_m(n) &\leqslant (p+1) \cdot 9^m < (p+1) \cdot 9^{p-1} \\ &< 9^p + C_p^1 \cdot 9^{p-1} < (9+1)^p = 10^p \leqslant n. \end{aligned}$$

This indicates that the terms of $\{n_k\}$ satisfies that if $n_k \geqslant 10^{m+1}$, then $n_{k+1} = S_m(n_k) < n_k$.

On the other hand, if positive integer $n < 10^{m+1}$, then

$$S_m(n) \leqslant (m+1) \cdot 9^m < (9+1)^{m+1} = 10^{m+1}.$$

We can get that if $n_k < 10^{m+1}$, then $n_{k+1} = S_m(n_k) < 10^{m+1}$ as well.

The discussion above indicates that when the subscript k is large enough, $n_k < 10^{m+1}$ must hold. Hence from some term on, every term of $\{n_k\}$ belongs to set $\{1, 2, \cdots, 10^{m+1} - 1\}$. That is, there exists $k_0 \in \mathbf{N}^*$ such that for any $k \geqslant k_0$, $1 \leqslant n_k \leqslant 10^{m+1} - 1$ holds. Combining this with pigeonhole principle, we can get that there exist $r, s \in \mathbf{N}^*$, $r > s \geqslant k_0$ such that $n_r = n_s$. By the definition of $\{n_k\}$, $n_k = n_{k+T}$ holds for $k \geqslant s$, where $T = r - s$ and $T \leqslant 10^{m+1} - 1$.

Hence $\{n_k\}$ is always a periodic sequence for any $n_0 \in \mathbf{N}^*$, with least positive period $\leqslant 10^{m+1} - 1$.

Therefore, (1) and (2) hold.

Example 6. First choose a positive integer a_0, and then $a_1 \in \{a_0 + 54, a_0 + 77\}$, and so on. When a_k is fixed, we can choose $a_{k+1} \in \{a_k + 54, a_k + 77\}$. In this way, we can get an infinite sequence $\{a_n\}$. Prove that there always exists a term whose last two number are same.

Proof. Discuss this question by a_n modulo 100. Let b_n denote the remainder when a_n is divided by 100. Equivalently, b_n is a double-digit number among 00, 01, $\cdots$, 99.

By the definition of $\{a_n\}$, for any $n \in \mathbf{N}^*$, $b_{n+1} \equiv b_n + 77$ or $b_n + 2 \times 77 \pmod{100}$ holds.

Since $(77, 100) = 1$, when j traverses the complete system of residues modulo 100, $77j$ traverses the complete system of residues modulo 100 as well. For $j = 0, 1, 2, \cdots, 99$, we arrange the remainder of $77j$ divided by 100 on a circle as shown on the right. Then from the structure of $\{b_n\}$, we can deduce that b_n and b_{n+1} are next to each other or separated by one number. Hence there must be one in any two adjacent numbers that is in $\{b_n\}$. Since 00 and 77 are adjacent, there must be $n \in \mathbf{N}^*$ such that $b_n = 00$ or 77, i.e., the last two digits of a_n are 00 or 77.

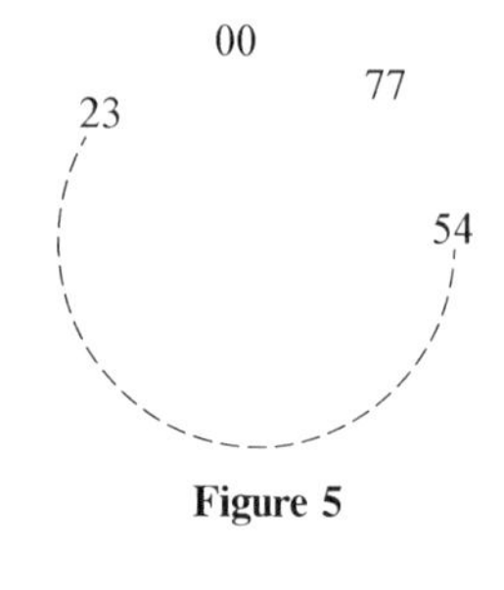

Figure 5

Therefore the proposition is true.

Explanation. Each term of $\{a_n\}$ has two choices and furthermore $\{a_n\}$ is not changing periodically under mod 100. But it does jump regularly and the method of combination helps solve the question smoothly.

Exercise Set 1

1. For any nonempty finite set, prove that we can arrange all of its subsets in a line, such that the difference between the numbers of elements of any adjacent subsets is 1.

2. Sequence $\{a_n\}$ satisfies that $a_0 = 0$, $a_n + a_{n-2} \geqslant 2a_{n-1}$, $n = 2, 3, \cdots$.

For any $n \in \mathbf{N}^*$ and $k \in \mathbf{Z}$, prove that $na_k \leqslant ka_n$ holds as long as $0 \leqslant k \leqslant n$.

3. Sequence $\{a_n\}$ of positive real numbers satisfying that $a_n^2 \leqslant a_n -$

a_{n+1}, $n = 1, 2, \cdots$. Prove that $a_n < \frac{1}{n}$ holds for any $n \in \mathbf{N}^*$.

4. Let real numbers $a_1, \cdots, a_n (n \geqslant 2)$ satisfy that $a_1 < a_2 < \cdots < a_n$. Prove that

$$a_1 a_2^4 + a_2 a_3^4 + \cdots + a_n a_1^4 \geqslant a_2 a_1^4 + a_3 a_2^4 + \cdots + a_n a_{n-1}^4 + a_1 a_n^4.$$

5. Let $a_1 = 1$, $a_2 = 2$, $a_{n+1} = \frac{a_n a_{n-1} + 1}{a_{n-1}}$, $n = 2, 3, \cdots$.

Prove that $a_n > \sqrt{2n}$ holds for any positive integer $n \geqslant 3$.

6. Let a be a positive real number. Prove that

$$\frac{1 + a^2 + a^4 + \cdots + a^{2n}}{a + a^3 + a^5 + \cdots + a^{2n-1}} \geqslant \frac{n+1}{n}$$

holds for any $n \in \mathbf{N}^*$.

7. Prove that

$$\lg(n!) > \frac{3n}{10}\left(\frac{1}{2} + \frac{1}{3} + \cdots + \frac{1}{n}\right)$$

holds for any $n \in \mathbf{N}^*$, $n \geqslant 2$.

8. Sequence $\{a_n\}$ of positive real numbers satisfying that $\sum_{j=1}^{n} a_j^3 = \left(\sum_{j=1}^{n} a_j\right)^2$ holds for any positive integer n. Prove that $a_n = n$ for any $n \in \mathbf{N}^*$.

9. Let $a_1, \cdots, a_n$ be n different positive integers. Prove that

$$a_1^2 + \cdots + a_n^2 \geqslant \frac{2n+1}{3}(a_1 + \cdots + a_n).$$

10. Sequence $\{a_n\}$ of real numbers satisfies that

(1) $a_1 = 2$, $a_2 = 500$, $a_3 = 2000$;

(2) $\frac{a_{n+2} + a_{n+1}}{a_{n+1} + a_{n-1}} = \frac{a_{n+1}}{a_{n-1}}$, $n = 2, 3, \cdots$.

Prove that each element of $\{a_n\}$ is an integer, and $2^n \mid a_n$ holds for any positive integer n.

11. Let k be a positive integer given, and sequence $\{a_n\}$ satisfy that

$$a_1 = k+1,\ a_{n+1} = a_n^2 - ka_n + k,\ n = 1, 2, \cdots.$$

Prove that a_m and a_n are relatively prime, for any positive integers $m \neq n$.

12. Sequence $\{a_n\}$ satisfies that $a_0 = 1$, $a_n = a_{n+1} + a_{\left[\frac{n}{3}\right]}$, $n = 1, 2, \cdots$.

For any prime number p no larger than 13, prove that there are infinitely many terms of $\{a_n\}$ that are multiples of p.

13. Denote $\{x\}$ the decimal part of x. Prove that

$$\sum_{k=1}^{n^2} \{\sqrt{k}\} \leqslant \frac{n^2 - 1}{2}$$

holds for any $n \in \mathbf{N}^*$.

14. Let m, $n \in \mathbf{N}^*$, and $S_m(n) = \sum_{k=1}^{n} \left[\sqrt[k^2]{k^m}\right]$. Prove that

$$S_m(n) \leqslant n + m\left(\sqrt[4]{2^m} - 1\right)$$

holds, where $[x]$ is the greatest integer which is no larger than x.

15. Let k be a positive integer given. Consider sequence $\{a_n\}$ satisfying that:

$$a_0 = 1,\ a_{n+1} = a_n + \left[\sqrt[k]{a_n}\right],\ n = 0, 1, 2, \cdots.$$

Find the set comprised of all the integer elements of $\left\{\sqrt[k]{a_n}\right\}$ for every k.

16. Sequence $\{x_n\}$ satisfies that $x_1 = \dfrac{1}{2}$, $x_n = \dfrac{2n-3}{2n} x_{n-1}$, $n = 2, 3, \cdots$.

Prove that $x_1 + x_2 + \cdots + x_n < 1$ holds for any $n \in \mathbf{N}^*$.

17. Sequence $\{f(n)\}$ satisfies that

$$f(1) = 2,\ f(n+1) = (f(n))^2 - f(n) + 1,\ n = 1, 2, 3, \cdots.$$

Prove that

$$1-\frac{1}{2^{2^{n-1}}}<\frac{1}{f(1)}+\frac{1}{f(2)}+\cdots+\frac{1}{f(n)}<1-\frac{1}{2^{2^n}}$$

holds for any integer $n \geqslant 1$.

18. Two sequences $x_1, x_2, \cdots$ and $y_1, y_2, \cdots$ of real numbers satisfy that

$$x_{n+1}=x_n+\sqrt{1+x_n^2},\ y_{n+1}=\frac{y_n}{1+\sqrt{1+y_n^2}},\ n \geqslant 1.$$

Prove that $2<x_n y_n<3$ holds for any $n>1$.

19. Sequence $\{a_n\}$ satisfies that $a_0=\frac{1}{2}$, $a_{n+1}=\frac{2a_n}{1+a_n^2}$, $n \geqslant 0$, while sequence $\{c_n\}$ satisfies that $c_0=4$, $c_{n+1}=c_n^2-2c_n+2$, $n \geqslant 0$.

Prove that $a_n=\frac{2c_0c_1\cdots c_{n-1}}{c_n}$ holds for any $n \geqslant 1$.

20. Sequence $\{a_n\}$ satisfies that $a_1=1$, $a_{n+1}=\frac{a_n}{n}+\frac{n}{a_n}$, $n \geqslant 1$.

Prove that $[a_n^2]=n$ holds for any $n \in \mathbf{N}^*$, $n \geqslant 4$.

21. Let a be an irrational number, and n be an integer greater than 1. Prove that $(a+\sqrt{a^2-1})^{\frac{1}{n}}+(a-\sqrt{a^2-1})^{\frac{1}{n}}$ is an irrational number.

22. Sequence of real numbers $\{a_n\}$ is defined as follows. $a_1=t$, $a_{n+1}=4a_n(1-a_n)$, $n=1, 2, \cdots$. How many different t are there satisfying that $a_{2011}=0$?

23. Sequence of real numbers $x_1, x_2, \cdots, x_{2011}$ satisfies that $|x_i-x_{i+1}| \leqslant 1$ for $i=1, 2, \cdots, 2010$. Find the maximal possible value of $\sum_{i=1}^{n}|x_i|-|\sum_{i=1}^{n}x_i|$.

24. Let $a_0, a_1, a_2, \cdots$ be an infinite sequence of positive real numbers. Prove that there exist infinitely many positive integers n such that $1+a_n>\sqrt[n]{2}a_{n-1}$.

25. For any $n \in \mathbf{N}$, the function $F: \mathbf{N} \to \mathbf{N}$ satisfies that:

(1) $F(4n)=F(2n)+F(n)$;

(2) $F(4n+2)=F(4n)+1$;

(3) $F(2n+1) = F(2n)+1$.

Prove that the number of integers n satisfying that $0 \leqslant n \leqslant 2^m$ and $F(4n) = F(3n)$ is $F(2^{m+1})$.

26. Function $f: \mathbf{N}^* \to \mathbf{N}^*$ is defined as follows. $f(1)=1$ and for any positive integer n,

$$f(n+1) = \begin{cases} f(n)+2, \text{ if } n = f(f(n)-n+1), \\ f(n)+1, \text{ other } n. \end{cases}$$

(1) Prove that $f(f(n)-n+1) \in \{n, n+1\}$ holds for any $n \in \mathbf{N}^*$.

(2) Find the expression of $f(n)$.

27. Sequence $\{a_n\}$ is defined as follows:

$$a_1 = 0,\ a_n = a_{\left[\frac{n}{2}\right]} + (-1)^{\frac{n(n+1)}{2}},\ n = 2, 3, \cdots.$$

For each $k \in \mathbf{N}$, find the number of subscripts n satisfying that $2^k \leqslant n < 2^{k+1}$ and $a_n = 0$.

28. Sequence of real numbers $\{a_n\}$ satisfies that

$$x_1 = 1,\ x_{n+1} = \begin{cases} x_n - 2, \text{ if } x_n - 2 > 0, \text{ and } x_n - 2 \notin \{x_1, \cdots, x_n\}. \\ x_n + 3, \text{ other cases.} \end{cases}$$

For any positive integer $k > 1$, prove that there exists a subscript n such that $x_{n+1} = x_n + 3 = k^2$.

29. Let n be a positive odd number, θ be a real number satisfying that $\frac{\pi}{\theta}$ is an irrational number. Let $a_k = \tan\left(\theta + \frac{k\pi}{n}\right)$, $k = 1, 2, \cdots, n$. Prove that $\frac{a_1 + a_2 + \cdots + a_n}{a_1 a_2 \cdots a_n}$ is an integer and find its value.

30. For any $n \in \mathbf{N}^*$, prove that there exists a polynomial $P(x)$ of degree n with integer coefficients whose leading coefficient is 1 such that $2\cos n\varphi = P(2\cos\varphi)$, where φ is any real number.

31. Let α and $\cos\alpha\pi$ be rational numbers. Find all possible values $\cos\alpha\pi$ can take on.

32. Do there exist infinitely many points on the unit circle such that the distance between any two points is a rational number?

33. Let n be a positive integer no less than 2. Find all polynomials with real coefficients

$$P(x)=a_n x^n+a_{n-1}x^{n-1}+\cdots+a_0,\ (a_n\neq 0),$$

such that $P(x)$ has n real roots that are all no more than 1, and satisfies that

$$a_0^2+a_1a_n=a_n^2+a_0a_{n-1}.$$

34. Let $P(x)$ be a polynomial with integer coefficients satisfying that $P(n)>n$ for any $n\in\mathbf{N}^*$ and that there exists at least one term in sequence

$$P(1),\ P(P(1)),\ P(P(P(1))),\ \cdots$$

that is a multiple of m for any $m\in\mathbf{N}^*$. Prove that $P(x)=x+1$.

35. Let $P(x)$ be a polynomial with real coefficients whose degree is an odd number. It satisfies that

$$P(x^2-1)=P(x)^2-1$$

for any $x\in\mathbf{R}$. Prove that $P(x)=x$.

36. Function $f:\mathbf{N}\rightarrow\mathbf{N}$ satisfies that

(1) $|f(x)-f(y)|\leqslant|x-y|$ holds, for any real number x, y.

(2) There exists positive integer k such that $f^{(k)}(0)=0$, where

$$f^{(1)}(x)=f(x),\ f^{(n+1)}(x)=f(f^{(n)}(x)),\ n=1,\ 2,\ \cdots.$$

Prove that $f(0)=0$ or $f(f(0))=0$ holds.

37. Sequence $\{p(n)\}$ satisfies that

$$p_1=2,\ p_2=5,\ p_{n+2}=2p_{n+1}+p_n,\ n=1,\ 2,\ \cdots.$$

Prove that

$$p_n=\sum\frac{(i+j+k)!}{i!j!k!}$$

holds for any positive integer n. Here, we take the sum among all nonnegative integer groups (i, j, k) satisfying that $i+j+k=2n$.

38. Let $A_n=\left\{1+\frac{\alpha_1}{\sqrt{2}}+\cdots+\frac{\alpha_n}{(\sqrt{2})^n}\,\middle|\,\alpha_i=1\text{ or }-1\right.$, where $i=1$,

$2, \cdots, n\Big\}$.

(1) Find the number of different elements of A_n, for every $n \in \mathbf{N}^*$.

(2) Find the sum of the product of any two different elements in A_n, for every $n \in \mathbf{N}^*$.

39. Sequence $\{x_n\}$ is defined as follows

$$x_0 = 4, \ x_1 = x_2 = 0, \ x_3 = 3,$$
$$x_{n+4} = x_n + x_{n+1}, \ n = 0, 1, 2, \cdots.$$

Prove that $p \mid x_p$ holds for any prime number p.

40. Find all sequences of positive integers $a_0, a_1, \cdots, a_n$ such that

(1) $\dfrac{a_0}{a_1} + \dfrac{a_1}{a_2} + \cdots + \dfrac{a_{n-1}}{a_n} = \dfrac{99}{100}$;

(2) $a_0 = 1$ and $(a_{k+1} - 1)a_{k-1} \geqslant a_k^2(a_k - 1)$, $k = 1, 2, \cdots, n-1$.

41. Sequence $\{y_n\}$ is defined as follows. $y_2 = y_3 = 1$, and

$$(n+1)(n-2)y_{n+1} = n(n^2 - n - 1)y_n - (n-1)^3 y_{n-1}, \ n = 3, 4, \cdots.$$

Prove that y_n is an integer if and only if n is prime.

42. Let $p > 3$ be a prime number, and $q = p^3$. Sequence $\{a_n\}$ is defined as follows.

$$a_n = \begin{cases} n, \ n = 0, 1, 2, \cdots, p-1, \\ a_{n-1} + a_{n-p}, \ n > p-1. \end{cases}$$

Find the remainder of a_q divided by p.

43. Let n be a positive integer no less than 2, and b_0 be an integer satisfying $2 \leqslant b_0 \leqslant 2n - 1$. Consider the sequence $\{b_i\}$ determined by

$$b_{i+1} = \begin{cases} 2b_i - 1, \text{ if } b_i \leqslant n, \\ 2b_i - 2n, \text{ if } b_i > n. \end{cases}$$

Let $p(b_0, n)$ denote the minimal subscript p satisfying that $b_p = b_0$.

(1) Let k be a positive integer, find the value of $p(2, 2^k)$ and $p(2, 2^k + 1)$.

(2) Prove that $p(b_0, n) \mid p(2, n)$ holds for any n and b_0.

44. Given a broken line starting from $(0, 0)$ and ending at $(1, 0)$

on a coordinate plane.

For any $n \in \mathbf{N}^*$, prove that there exist two points on the broken line with same ordinate, whose abscissas differ $\frac{1}{n}$ from each other.

45. There are a black box and n white boxes marked by 1, 2, $\cdots$, n and n white balls placed in the white boxes. The following operations are allowed: If there are exactly k white balls in the box labeled by k, then take all the k balls out, and place one ball respectively in the black box and white boxes labeled by 1, 2, $\cdots$, $k-1$. For any $n \in \mathbf{N}^*$, prove that there exists only one way of placement such that the n balls are placed in the white boxes at the beginning, and are placed in the black box after limited times of operations.

46. Let R_0 be an array with n elements, whose elements belong to $\{A, B, C\}$. Define sequences R_0, R_1, R_2, $\cdots$ as follows. If $R_j = (x_1, \cdots, x_n)$, then $R_{j+1} = (y_1, y_2, \cdots, y_n)$, where

$$y_i = \begin{cases} x_i, \text{ if } x_i = x_{i+1}, \\ \{A, B, C\} \setminus \{x_i, x_{i+1}\}, \text{ if } x_i \neq x_{i+1}, \end{cases}$$

and $x_{n+1} = x_1$. For instance, if $R_0 = (A, A, B, C)$, then $R_1 = (A, C, A, B)$, $R_2 = (B, B, C, C)$, $\cdots$.

(1) Find all $n \in \mathbf{N}^*$ such that there exists $m \in \mathbf{N}^*$ satisfying that $R_m = R_0$ for any R_0.

(2) Find the minimal positive integer m satisfying (1) for $n = 3^k (k \in \mathbf{N}^*)$.

Chapter 2 Selected Topical Discussions

9 The Fibonacci Sequence

The Fibonacci sequence $\{F_n\}$ is defined as follows.

$$F_1 = F_2 = 1, \ F_{n+2} = F_{n+1} + F_n, \ n = 1, 2, \cdots.$$

This is a well-known sequence. There are endless discussions about it. Many interesting and profound conclusions are drawn. Some of them are displayed by examples here.

Example 1. For any $m, n \in \mathbf{N}^*$, prove that $(F_m, F_n) = F_{(m, n)}$, i.e. the greatest common factor of the terms in the Fibonacci sequence can be transformed into its subscripts.

Proof. It is obviously true when $m = n$. We consider the situation when $m \neq n$. Without loss of generality, let $m > n$.

Applying the recurrence formula of the Fibonacci sequence, we have that

$$\begin{aligned} F_m &= F_{m-1} + F_{m-2} = F_2 F_{m-1} + F_1 F_{m-2} \\ &= F_2(F_{m-2} + F_{m-3}) + F_1 F_{m-2} \\ &= (F_2 + F_1) F_{m-2} + F_2 F_{m-3} \\ &= F_3 F_{m-2} + F_2 F_{m-3} \\ &= \cdots = F_n F_{m-n+1} + F_{n-1} F_{m-n}. \end{aligned}$$

Therefore, $(F_m, F_n) = (F_{n-1} F_{m-n}, F_n) = (F_{m-n}, F_n)$. (Here we utilize the fact that $(F_{n-1}, F_n) = 1$, which can be proved by mathematical induction on n. Details are left to readers.)

From the result above, we continue the discussion by substituting (m, n) by $(m-n, n)$, which implies that solving the greatest common factor of F_m and F_n is essentially taking a division algorithm on subscripts m and n. So, $(F_m, F_n) = F_{(m, n)}$.

Explanation. The following proposition can be proved by the result of this example. If F_n is a prime number, then $n = 4$ or n is a prime number.

In fact, if $n \neq 4$ and n is not a prime number, then n can be represented as $n = pq$, $2 \leqslant p \leqslant q$ and $q \geqslant 3$. Now $(F_n, F_q) = F_{(n, q)} = F_q$. Meanwhile, $F_q \geqslant 2$, $F_n > F_q$, which derives that F_n is a composite number.

Example 2. Prove that every positive integer m can be expressed in the following form uniquely.

$$\begin{aligned} m &= (a_n a_{n-1} \cdots a_2)_F \\ &= a_n F_n + a_{n-1} F_{n-1} + \cdots + a_2 F_2. \end{aligned} \quad ①$$

Here $a_i = 0$ or 1, $a_n = 1$. Also, there is no subscript satisfying $2 \leqslant i \leqslant n-1$ such that $a_i = a_{i+1} = 1$. In the formula, F_i is the ith term of the Fibonacci sequence.

Proof. The positive integer expression by formula ① is regarded as F-expression of m. It is similar to the binary system. This result is the famous Zerkendorf Theorem.

It will be proved by induction on m .

When $m = 1$, $m = F_2$. The proposition is proved. Now suppose the proposition is true for any positive integer k less than m.

There is a unique $n \in \mathbf{N}^*$ such that $F_n \leqslant m < F_{n+1}$. If $m - F_n = 0$, then m is expressed in the form of ①. Otherwise, if $m - F_n > 0$, from the inductive hypothesis, $m - F_n$ is expressed in the form of ①. Let

$$m - F_n = (a_l a_{l-1} \cdots a_2)_F = a_l F_l + \cdots + a_2 F_2.$$

Here if $a_l = 1$, then $m = F_n + a_l F_l + \cdots + a_2 F_2$. Now if $l \geqslant n-1$, then $m \geqslant F_n + F_{n-1} = F_{n+1}$, which is contradictory. So $l \leqslant n-2$.

Positive integer m can be expressed in the form of ①.

The following part demonstrates the uniqueness of expression ① of m.

In fact, if

$$m = (a_n \cdots a_2)_F = (b_l \cdots b_2)_F, \tag{②}$$

where $a_n = b_l = 1$. and $n \geqslant l$.

If $n > l$, there is no subscript $1 \leqslant i \leqslant l - 1$ satisfying $b_i = b_{i+1} = 1$. Noting the definition of $\{F_n\}$ we have

$$\begin{aligned}(b_l \cdots b_2)_F &\leqslant \begin{cases} F_l + F_{l-2} + \cdots + F_3, & m \text{ is even}, \\ F_l + F_{l-2} + \cdots + F_4 + F_2, & m \text{ is odd}, \end{cases} \\ &< \begin{cases} F_l + F_{l-2} + \cdots + F_3 + F_2 = F_{l+1}, & m \text{ is even}, \\ F_l + F_{l-2} + \cdots + F_4 + F_2 + F_1 = F_{l+1}, & m \text{ is odd}. \end{cases}\end{aligned}$$

Hence $(b_l \cdots b_2)_F < F_{l+1} \leqslant F_n$. Equal signs cannot be taken at the same time in ②.

Therefore $n = l$ and then $m - F_n$ has two expressions, which is contrary to the inductive hypothesis. There is a unique way to express m.

In summary, by the second form of mathematical induction, the proposition is true.

Example 3. It is well-known that the product of n consecutive positive integers is a multiple of that of the first n positive integers (namely $n!$). The Fibonacci sequence has a similar property. For any $k \in \mathbf{N}^*$, please prove that the product of any k consecutive terms of $\{F_n\}$ is a multiple of that of the first k terms.

Proof. Notation is imported. $[n]! = F_1 F_2 \cdots F_n$, $n = 1, 2, \cdots$. Set $[0]! = 1$. And write

$$R(m, n) = \frac{[m+n]!}{[m]! \cdot [n]!}, \quad m, n \in \mathbf{N}.$$

To prove the proposition, we need only prove $R(m, n) \in \mathbf{N}^*$, for any $m, n \in \mathbf{N}^*$.

Taking a similar derivation as in example 1, we know

$$F_{m+n} = F_2F_{m+n-1} + F_1F_{m+n-2} = \cdots = F_mF_{n+1} + F_{m-1}F_n.$$

Therefore, we have

$$\begin{aligned} R(m, n) &= \frac{F_{m+n} \cdot [m+n-1]!}{[m]! \cdot [n]!} = \frac{F_{m+n} \cdot [m+n-1]!}{F_m \cdot F_n \cdot [m-1]! \cdot [n-1]!} \\ &= F_{n+1} \cdot \frac{[m+n-1]!}{[m-1]! \cdot [n]!} + F_{m-1} \cdot \frac{[m+n-1]!}{[m]! \cdot [n-1]!} \\ &= F_{n+1} \cdot R(m-1, n) + F_{m-1} \cdot R(m, n-1). \end{aligned}$$

The formula above holds for $m, n \in \mathbf{N}^*$. Noting the initial condition $R(0, n) = R(m, 0) = 1$ (it holds for any $m, n \in \mathbf{N}^*$) and by mathematical induction, we can prove that $R(m, n)$ is a positive integer.

Hence, the proposition is true.

Example 4. Let $f(x) = \frac{1}{x+1}$ $(x > 0)$. Prove that

(1) For any positive integer n,

$$g_n(x) = x + f(x) + f(f(x)) + \cdots + \underbrace{f(f(\cdots f(x)))}_{n \text{ iterations}}$$

is an increasing function on $(0, +\infty)$;

(2) $g_n(1) = \frac{F_1}{F_2} + \frac{F_2}{F_3} + \cdots + \frac{F_{n+1}}{F_{n+2}}$, where $\{F_n\}$ is the Fibonacci sequence.

Proof. For convenient formulation, we notate

$$f^{(n)}(x) = \underbrace{f(f(\cdots f(x)))}_{n \text{ iterations}}.$$

This function iteration problem is discussed locally.

(1) It is familiar that $y = x + \frac{1}{x}$ is monotonically increasing on $(1, +\infty)$. Accordingly,

$$h(x) = x + f(x) = x + \frac{1}{1+x} = (1+x) + \frac{1}{1+x} - 1$$

is monotonically increasing on $(0, +\infty)$.

Noting that

$$f(f(x)) = \frac{1}{1+f(x)} = \frac{1}{1+\frac{1}{1+x}} = \frac{1+x}{2+x} = 1 - \frac{1}{2+x}$$

is an increasing function on $(0, +\infty)$, we have that the function $f^{(2k)}(x)$ is an increasing function on $(0, +\infty)$ for any $k \in \mathbf{N}^*$. Since $h(x)$ is increasing on $(0, +\infty)$, $f^{(2k)}(x) + f^{(2k+1)}(x)$ is also increasing on $(0, +\infty)$.

With the conclusions above, when n is odd, function

$$g_n(x) = (x + f(x)) + (f^{(2)}(x) + f^{(3)}(x)) + \cdots + (f^{(n-1)}(x) + f^{(n)}(x))$$

is the sum of $\frac{n+1}{2}$ increasing functions on $(0, +\infty)$.

When n is even, $f^{(n)}(x)$ and $g_n(x) - f^{(n)}(x)$ are both increasing functions on $(0, +\infty)$. Hence, $g_n(x)$ is also an increasing function on $(0, +\infty)$.

Therefore for any $n \in \mathbf{N}^*$, $g_n(x)$ is an increasing function on $(0, +\infty)$.

(2) With the definition of $g_n(x)$, we need only to prove that $f^{(n)}(1) = \frac{F_{n+1}}{F_{n+2}}$, for any $n \in \mathbf{N}^*$ (here $f^{(0)}(x) = x$).

Utilizing that $1 = \frac{F_1}{F_2}$ and $f(1) = \frac{1}{2} = \frac{F_2}{F_3}$, we get that the statement is true when $n = 0, 1$. Now suppose that $f^{(n)}(1) = \frac{F_{n+1}}{F_{n+2}}$ (i.e., the proposition is true for n). Since $f^{(n+1)}(x) = \frac{1}{1+f^{(n)}(x)}$, we have $f^{(n+1)}(1) = \frac{1}{1+f^{(n)}(1)}$. Hence

$$f^{(n+1)}(x) = \frac{1}{1+\frac{F_{n+1}}{F_{n+2}}} = \frac{F_{n+2}}{F_{n+2}+F_{n+1}} = \frac{F_{n+2}}{F_{n+3}}.$$

Therefore (2) holds.

Example 5. Consider sequence $\{x_n\}$: $x_1 = a$, $x_2 = b$, $x_{n+2} = x_{n+1} + x_n$, $n = 1, 2, 3, \cdots$, where a, b are real numbers. If there exist positive integers k and m, $k \neq m$, such that $x_k = x_m = c$, then real number c is called "Double Value". Prove that there are real numbers a, b such that at least 2,000 different "Double Values" exist. Moreover, prove that we cannot find a, b such that infinitely many "Double Values" exist.

Proof. A sequence with 2,000 different "Double Values" is established with the Fibonacci sequence.

The idea is to extend the subscripts of $\{F_n\}$ to negative integers corresponding to the original recurrence relation. We have

$$\begin{aligned}
F_0 &= F_2 - F_1 = 0, \\
F_{-1} &= F_1 - F_0 = 1 = F_1, \\
F_{-2} &= F_0 - F_{-1} = -1 = -F_2, \\
F_{-3} &= F_{-1} - F_{-2} = 2 = F_3,
\end{aligned}$$

and so on. Now we know $F_{-2m} = -F_{2m}$, $F_{-(2m+1)} = F_{2m+1}$, $m = 1, 2, \cdots$. Hence for any $m \in \mathbf{N}^*$, let $a = F_{2m+1}$, $b = F_{2m}$. Then the sequence $\{x_n\}$ is displayed as

$$F_{2m+1},\ -F_{2m},\ F_{2m-1},\ -F_{2m-2},\ \cdots,\ -F_2,\ F_1,\ F_0,$$
$$F_1,\ F_2,\ \cdots,\ F_{2m-1},\ F_{2m},\ F_{2m+1},\ \cdots.$$

Terms F_1, F_3, $\cdots$, F_{2m+1} are all 'Double Values' in sequence $\{x_n\}$. Specifically, the required sequence $\{x_n\}$ is obtained by taking $m = 1999$.

On the other hand, if we can find a, b such that infinitely many 'Double Values' appear in sequence $\{x_n\}$. Then any two adjacent terms in $\{x_n\}$ have opposite signs. (Otherwise, the sequences become strictly increasing (or strictly decreasing) starting from the next term. There can't be infinitely many different "Double Values".)

Note that the characteristic equation of $\{x_n\}$ (also the characteristic equation of the Fibonacci sequence) is $\lambda^2 = \lambda + 1$ which have two distinct real solutions. Hence we assume that

$$x_n = A \cdot \left(\frac{1+\sqrt{5}}{2}\right)^n + B \cdot \left(\frac{1-\sqrt{5}}{2}\right)^n, \ n = 1, 2, \cdots.$$

Because $\left|\frac{1-\sqrt{5}}{2}\right| < 1$ and $\left|\frac{1+\sqrt{5}}{2}\right| > 1$, if $A > 0$ then $x_n > 0$ when n is sufficiently large. So two positive adjacent terms occur. Similarly, if $A < 0$, then two negative adjacent terms occur in $\{x_n\}$. Both situations lead to contradictions. Therefore $A = 0$. Then $x_n = B \cdot \left(\frac{1-\sqrt{5}}{2}\right)^n$. Noting that $\left|\frac{1-\sqrt{5}}{2}\right| < 1$, we have that the sequence $\{|x_n|\}$ is monotonically decreasing. There is no "Double Value" when $B \neq 0$. When $B = 0$, however, there is only one "Double Value".

In conclusion, the proposition is true.

Explanation. The general formula of the Fibonacci sequence can be solved by its characteristic equation and initial values.

$$F_n = \frac{1}{\sqrt{5}}\left(\frac{1+\sqrt{5}}{2}\right)^n - \frac{1}{\sqrt{5}}\left(\frac{1-\sqrt{5}}{2}\right)^n, \ n = 1, 2, \cdots.$$

But the recurrence relation is more useful than the general formula when solving practical problems.

Example 6. Arrange terms in the Fibonacci sequence in order: 1, 1, 2, 3, 5, 8, $\cdots$. Sort all the Twin Primes (if p and $p+2$ are both primes, then p and $p+2$ are Twin Primes) by size: 5, 7, 11, 13, 17, 19, 29, 31,$\cdots$. Find the positive integers that appear in both sequences.

Solution. Comparing several terms at the beginning of the two sequences, we observe that only number 3, 5 and 13 show up in both sequences. We may guess that these positive integers are all what is required.

Due to the difficulty of understanding the patterns of Twin Prime sequence, in order to testify the conjecture above, we should begin with the properties of the Fibonacci sequence. If n is fairly large, either F_n is composite or $F_n \pm 2$ are both composite numbers, then F_n is not in the Twin Prime sequence. With this idea, we need to tell some properties of the Fibonacci sequence first.

List several first terms of the Fibonacci sequence.

n	1	2	3	4	5	6	7	8	9	10	11	12	13	14	15	⋯
F_n	1	1	2	3	5	8	13	21	34	55	89	144	233	377	610	⋯

We find that F_{2n} (when $n \geqslant 3$) is a composite number. $F_{2n} \pm 2$ (when $n \geqslant 4$) are composite numbers too. Besides, there are some formulas as follows.

(1) $F_{2n} = F_n(F_{n+1} + F_{n-1})$, with $F_0 = 0$;

(2) $F_{4n+1} + 2 = F_{2n-1}(F_{2n+1} + F_{2n+3})$;

(3) $F_{4n+1} - 2 = F_{2n+2}(F_{2n-2} + F_{2n})$;

(4) $F_{4n+3} + 2 = F_{2n+3}(F_{2n+1} + F_{2n-1})$;

(5) $F_{4n+3} - 2 = F_{2n}(F_{2n+2} + F_{2n+4})$.

Note that if these five formulas are proved, then only 3, 5 and 13 appear in both sequences.

Now we prove (1)-(5) by mathematical induction.

When $n = 1$, it is obvious that (1)-(5) hold by data listed in the previous table. Now assume (1)-(5) are proved for integers no larger than n. By the recurrence formula of the Fibonacci sequence, for $n+1$, we have

$$\begin{aligned}
F_{4n+2} &= F_{4n+1} + F_{4n} = F_{4n+1} + F_{4n-1} + F_{4n-2} \\
&= (F_{4n+1} + 2) + (F_{4n-1} - 2) + F_{4n-2} \\
&= F_{2n-1}(F_{2n+1} + F_{2n+3}) + F_{2n-2}(F_{2n} + F_{2n+2}) + F_{2n-1}(F_{2n-2} + F_{2n}) \\
&= F_{2n+1}F_{2n-1} + F_{2n-1}F_{2n+3} + F_{2n-2}(F_{2n} + F_{2n-1}) \\
&\quad + (F_{2n-2}F_{2n+2} + F_{2n-1}F_{2n}) \\
&= F_{2n+1}F_{2n-1} + F_{2n-2}F_{2n+1} + F_{2n-1}(F_{2n+3} + F_{2n}) + F_{2n-2}F_{2n+2} \\
&= F_{2n+1}F_{2n} + 2F_{2n-1}F_{2n+2} + (F_{2n} - F_{2n-1})F_{2n+2} \\
&= F_{2n+1}F_{2n} + F_{2n+2}(F_{2n-1} + F_{2n}) \\
&= F_{2n+1}(F_{2n} + F_{2n+2}),
\end{aligned}$$

i.e.

$$F_{2(2n+1)} = F_{2n+1}(F_{2n+1-1} + F_{2n+1+1}). \qquad ①$$

Similarly, we can prove

$$F_{4n+4} = F_{2n+2}(F_{2n+1} + F_{2n+3}),$$

i.e.,

$$F_{2(2n+2)} = F_{2n+2}(F_{2n+2-1} + F_{2n+2+1}). \qquad ②$$

From ① and ② we know (1) holds for $2n+1$ and $2n+2$. Therefore it is true for all $n \in \mathbf{N}^*$.

$$\begin{aligned} F_{4n+5} + 2 &= F_{4n+4} + (F_{4n+3} + 2) \\ &= F_{2n+2}(F_{2n+1} + F_{2n+3}) + F_{2n+3}(F_{2n+1} + F_{2n-1}) \\ &= F_{2n+1}(F_{2n+2} + F_{2n+3}) + F_{2n+3}(F_{2n+2} + F_{2n-1}) \\ &= F_{2n+1}F_{2n+4} + 2F_{2n+3}F_{2n+1} \\ &= F_{2n+1}(F_{2n+3} + F_{2n+5}), \end{aligned}$$

i.e. (2) is testified for $n+1$. Similarly, (3), (4), (5) hold for $n+1$. (Details are left to readers.)

Generalizing all above, (1)-(5) are true for all $n \in \mathbf{N}^*$. Only 3, 5 and 13 appear in both sequences.

Explanation. With example 1, we can see that F_{2n} is a composite number when $n \geqslant 3$. The result here is more powerful.

10 Several Proofs of AM-GM Inequality

From this section, we are going to discuss some other forms and techniques of applying mathematical induction.

AM-GM Inequality Let $a_1, a_2, \cdots, a_n$ be n positive real numbers. Then

$$\frac{a_1 + a_2 + \cdots + a_n}{n} \geqslant \sqrt[n]{a_1 a_2 \cdots a_n}. \qquad ①$$

$\frac{1}{n}(a_1 + \cdots + a_n)$ is called the arithmetic mean of $a_1, a_2, \cdots, a_n$, while $\sqrt[n]{a_1 a_2 \cdots a_n}$ is called the geometric mean of $a_1, a_2, \cdots, a_n$.

Proof 1. When $n=1$, it is obvious that ① holds; when $n=2$, ① is equivalent to $a_1+a_2 \geqslant 2\sqrt{a_1a_2}$, i.e. $(\sqrt{a_1}-\sqrt{a_2})^2 \geqslant 0$. Hence ① holds.

Now we suppose that ① holds for $n(\geqslant 2)$ and consider the case of $n+1$.

Let $A=\frac{1}{n+1}(a_1+\cdots+a_{n+1})$. Then by inductive hypothesis, we have

$$\begin{aligned}
&\frac{1}{2n}(a_1+\cdots+a_{n+1}+\underbrace{A+\cdots+A}_{n-1 \text{ items}})\\
&=\frac{1}{2n}(a_1+\cdots+a_n)+\frac{1}{2n}(a_{n+1}+\underbrace{A+\cdots+A}_{n-1 \text{ items}})\\
&\geqslant \frac{1}{2}(\sqrt[n]{a_1\cdots a_n}+\sqrt[n]{a_{n+1}\underbrace{A\cdots A}_{n-1 \text{ items}}})\\
&\geqslant \sqrt{\sqrt[n]{a_1\cdots a_n}\cdot\sqrt[n]{a_{n+1}\cdot A^{n-1}}}\\
&=\sqrt[2n]{a_1\cdots a_{n+1}A^{n-1}}.
\end{aligned}$$

Noting that $a_1+\cdots+a_{n+1}=(n+1)A$, we can deduce that

$$\frac{1}{2n}(a_1+\cdots+a_{n+1}+\underbrace{A+\cdots+A}_{n-1 \text{ items}})=\frac{1}{2n}((n+1)A+(n-1)A)=A.$$

Therefore

$$A \geqslant \sqrt[2n]{a_1\cdots a_{n+1}A^{n-1}},$$

moreover, $A^{n+1}\geqslant a_1\cdots a_{n+1}$, then we get

$$\frac{1}{n+1}(a_1+\cdots+a_{n+1})\geqslant \sqrt[n+1]{a_1\cdots a_{n+1}}.$$

Hence ① holds for $n+1$.

Thus, for any $n\in\mathbf{N}^*$, inequality ① holds.

Explanation. This is a proof of ① given by the first form of mathematical induction, which is quite skillful.

Proof 2. Since ① holds for $n=2$ (the proof is same to Proof 1), it can be proved easily by mathematical induction that ① holds for all $n = 2^k (k \in \mathbf{N}^*)$.

In fact, if ① holds for 2^k, then for $n = 2^{k+1}$, there holds

$$\begin{aligned}\frac{1}{2^{k+1}}(a_1 + \cdots + a_{2^{k+1}}) &= \frac{1}{2}\left(\frac{1}{2^k}(a_1 + \cdots + a_{2^k}) + \frac{1}{2^k}(a_{2^k+1} + \cdots + a_{2^{k+1}})\right)\\ &\geqslant \frac{1}{2}\left(\sqrt[2^k]{a_1 \cdots a_{2^k}} + \sqrt[2^k]{a_{2^k+1} \cdots a_{2^{k+1}}}\right)\\ &\geqslant \sqrt{\sqrt[2^k]{a_1 \cdots a_{2^k}} \cdot \sqrt[2^k]{a_{2^k+1} \cdots a_{2^{k+1}}}}\\ &= \sqrt[2^{k+1}]{a_1 a_2 \cdots a_{2^{k+1}}}.\end{aligned}$$

That is, ① holds for all $n = 2^k$, $k = 1, 2, \cdots$.

Let's discuss the case of n. For any $n \in \mathbf{N}^*$, take $k \in \mathbf{N}$ satisfying that $2^k \leqslant n < 2^{k+1}$ and denote $A = \frac{1}{n}(a_1 + \cdots + a_n)$. From the previous result, we can deduce that

$$\frac{1}{2^{k+1}}(a_1 + \cdots + a_n + \underbrace{A + \cdots + A}_{2^{k+1} - n \text{ items}}) \geqslant \sqrt[2^{k+1}]{a_1 \cdots a_n \underbrace{A \cdots A}_{2^{k+1} - n \text{ items}}}.$$

Noting that $a_1 + \cdots + a_n = nA$, we have

$$A \geqslant \sqrt[2^{k+1}]{a_1 \cdots a_n A^{2^{k+1} - n}}.$$

Moreover, $A^n \geqslant a_1 \cdots a_n$. Hence ① holds for n.

Explanation. This proof comes quite naturally. Both of the proofs needs to piece together terms.

Proof 3. It can be inferred from the previous proof that ① holds for $n = 2^k$, which shows that there exist infinitely many positive integers n such that ① holds.

Now suppose that ① holds for $n + 1$. As to the case of n, denoting $A = \frac{1}{n}(a_1 + \cdots + a_n)$, we have

$$\frac{1}{n+1}(a_1 + \cdots + a_n + A) \geqslant \sqrt[n+1]{a_1 \cdots a_n \cdot A}.$$

Noting that $a_1 + \cdots + a_n = nA$, we get $A \geqslant \sqrt[n+1]{a_1 \cdots a_n \cdot A}$. And thus, $A \geqslant \sqrt[n]{a_1 \cdots a_n}$, which implies that ① holds for n.

Hence ① holds for any $n \in \mathbf{N}^*$.

Explanation. Here, we apply the idea of patching up a hole, which is a basic application of inverse mathematical induction.

Inverse induction is also called backward induction, whose basic structure is as follows: suppose a proposition (or property) $P(n)$ about (of) positive integer n satisfy the following conditions

(1) $P(n)$ is true for infinitely many positive integers n.

(2) It can be inferred from the validity of $P(n+1)$ that $P(n)$ is true.

Then for any $n \in \mathbf{N}^*$, P(n) is true.

Proof. Let's prove by contradiction.

If there exists $m \in \mathbf{N}^*$ such that $P(m)$ is not true. We can prove by mathematical induction that for any $n \geqslant m$, $P(n)$ is not true (then there can be only a finite number of $n \in \mathbf{N}^*$ such that $P(n)$ is true, which is contrary to condition (1)).

In fact, $P(m)$ is not true according to the assumption.

Now suppose that $P(n)(n \geqslant m)$ is not true. Then it can be inferred from (2) that $P(n+1)$ is not true (by the contrapositive of (2)).

Hence by mathematical induction, we have that for any $n \geqslant m$, $P(n)$ is not true, which is contradictory. Therefore inverse induction holds.

Both of the two latter proofs of AM-GM inequality proved first that the proposition is true for infinitely many $n \in \mathbf{N}^*$. Then they prove that the proposition is true for every $n \in \mathbf{N}^*$. This idea is applied in many cases.

AM-GM inequality may be the theorem with most proofs in mathematics. Here we only give some most common methods with mathematical induction. These ideas and methods can be applied to other questions.

Example 1. Let function $f: \mathbf{N}^* \to [1, +\infty)$ satisfy:

(1) $f(2) = 2$;

(2) for any $m, n \in \mathbf{N}^*$, $f(mn) = f(m)f(n)$;

(3) when $m < n$, $f(m) < f(n)$.

For any positive integer n, prove that $f(n) = n$ holds.

Proof. It can be inferred from condition (1) and (2) that $f(1) = 1$. Now suppose that $f(2^k) = 2^k$, $k \in \mathbf{N}$. Then $f(2^{k+1}) = f(2^k)f(2) = 2^k \times 2 = 2^{k+1}$. Hence for any $k \in \mathbf{N}$, $f(2^k) = 2^k$.

Now let's discuss the value of $f(n)$. Set $f(n) = l$. Then by (2) and mathematical induction, we have $f(n^m) = l^m$, for any $m \in \mathbf{N}^*$.

Letting $2^k \leqslant n^m < 2^{k+1}$, by (3), we get $f(2^k) \leqslant f(n^m) < f(2^{k+1})$. Therefore it can be deduced from the previous result that $2^k \leqslant l^m < 2^{k+1}$. Noting that $2^k \leqslant n^m < 2^{k+1}$, we have

$$\frac{1}{2} < \left(\frac{n}{l}\right)^m < 2. \qquad ①$$

This inequality holds for any $m \in \mathbf{N}^*$.

If $n > l$, let's take $m > \frac{l}{n - l}$. Then

$$\left(\frac{n}{l}\right)^m = \left(1 + \frac{n-l}{l}\right)^m \geqslant 1 + m \cdot \frac{n-l}{l} > 2,$$

which is contrary to ①. Similarly, if $n < l$, we can take $m > \frac{n}{l - n}$. Then $\left(\frac{l}{n}\right)^m > 2$, i.e., $\left(\frac{n}{l}\right)^m < \frac{1}{2}$, which is also contrary to ①. Therefore we can only have $n = l$.

Generalizing all above, for any $n \in \mathbf{N}^*$, $f(n) = n$.

Explanation. If function f is a mapping from $\mathbf{N}^*$ to $\mathbf{N}^*$, then the question is simpler, and it is left to readers.

Similarly, this method can be used to prove the famous Jensen's Inequality.

Example 2. Find all functions $f: \mathbf{N}^* \to \mathbf{N}^*$, satisfying that for any m, $n \in \mathbf{N}^*$,

$$f(m)^2 + f(n) \mid (m^2 + n)^2. \tag{①}$$

Solution. Let f be a function satisfying the conditions. Set $m = n = 1$ in ①. Then $(f(1)^2 + f(1)) \mid 4$. That is, $f(1)(f(1)+1) \mid 4$. If $f(1) \geqslant 2$, then $f(1)(f(1)+1) \geqslant 6$. Hence there can only be $f(1) = 1$.

For any prime number p, let's first prove that

$$f(p-1) = p-1. \tag{②}$$

In fact, for any prime number p, set $m = 1$, $n = p-1$ in ①. Then $(f(1)+f(p-1)) \mid p^2$, i.e. $(1+f(p-1)) \mid p^2$. Hence $f(p-1)+1 = p$ or p^2. If the former one is true, then ② holds; if the latter, i.e., $f(p-1) = p^2-1$, then set $m = p-1$, $n = 1$ in ①. We have $(f(p-1)^2 + f(1)) \mid ((p-1)^2+1)^2$, i.e. $((p^2-1)^2+1) \mid ((p-1)^2+1)^2$. However

$$\begin{aligned}((p-1)^2+1)^2 &\leqslant ((p-1)^2+(p-1))^2 \\ &= p^2(p-1)^2 < (p+1)^2(p-1)^2+1 \\ &= (p^2-1)^2+1,\end{aligned}$$

which is contradictory. Therefore ② holds.

Now for any $n \in \mathbf{N}^*$, let's prove that $f(n) = n$.

In fact, for any positive integer n, we can take $k \in \mathbf{N}^*$ such that $k+1$ is a prime number (there are infinitely many k of this kind). Set $m = k$ in ①. Combing with ②, we have

$$(k^2 + f(n)) \mid (k^2+n)^2. \tag{③}$$

Note that

$$\begin{aligned}(k^2+n)^2 &= (k^2+f(n)+n-f(n))^2 \\ &= A(k^2+f(n)) + (n-f(n))^2,\end{aligned}$$

where A is an integer. By ③, we have

$$(k^2+f(n)) \mid (n-f(n))^2.$$

This formula shows that $(n-f(n))^2$ is divisible by infinitely many positive integers (since there are infinite ways to take k). Hence $(n -$

$f(n))^2 = 0$, i.e., $f(n) = n$.

In conclusion, there is only one function satisfying the condition, $f(n) = n$.

Explanation. Essentially, it suffices to prove that $f(k) = k$ holds, for infinitely many $k \in \mathbf{N}^*$. Then patch up other holes. We break through the question from ② because we expect the factors of the dividend as few as possible. This technique is often applied in the theory of divisibility.

Example 3. Find all of the functions $f: \mathbf{N}^* \to \mathbf{N}^*$ satisfying that for any $n \in \mathbf{N}^*$ and prime number p,

$$f(n)^p \equiv n \pmod{f(p)}. \qquad ①$$

Solution. For any prime number p, taking $n = p$ in ①, we have

$$p \equiv f(p)^p \equiv 0 \pmod{f(p)}.$$

Hence $f(p) \mid p$ and then $f(p) = 1$ or p.

Now denote $S = \{p \mid p \text{ is prime}, f(p) = p\}$. The question can be divided into three cases:

Case 1. S is an infinite set. We can prove, by the method in the previous example, that for any $n \in \mathbf{N}^*$, $f(n) = n$.

In fact, there exist infinitely many prime numbers p such that $f(p) = p$. Hence for any $n \in \mathbf{N}^*$, there exist infinitely many prime numbers p such that $n \equiv f(n)^p \pmod{p}$. By Fermat's little theorem, we have $f(n)^p \equiv f(n) \pmod{p}$. Therefore $f(n) \equiv n \pmod{p}$, which shows that $f(n) - n$ is a multiple of infinitely many prime numbers. So $f(n) = n$.

Case 2. S is empty. Then for any prime number p, $f(p) = 1$. At this time, for the rest of positive integers n, $f(n)$ can take any positive integer (such that ① holds).

Case 3. S is a nonempty finite set.

Let p be the greatest prime number in S. If $p \geqslant 3$, we will prove this leads to a contradiction. Therefore $S = \{2\}$.

Since p is maximal, for any prime number $q > p$, $f(q) = 1$. From ①, it can be inferred that $q \equiv f(q)^p \equiv 1 \pmod{p}$, i.e., $q \equiv 1 \pmod{p}$.

Now let Q be the product of all odd prime numbers no greater than p. Then each prime factor of $Q+2$ is greater than p (note that we applied that $p \geqslant 3$ here). Then combining with the result above, we have $Q+2 \equiv 1 \pmod{p}$, leading to $p \mid Q+1$, which is contrary to the fact that $p \mid Q$.

The discussion above shows that $S = \{2\}$. So $f(2) = 2$. For odd prime number p, $f(p) = 1$. By ①, we only need to prove that $f(n)^2 \equiv n \pmod 2$. Hence for other positive integer n, $f(n)$ suffices to take on a positive integer with the same odevity with n.

Checking directly, we can find that each function in the three cases satisfies the conditions. They are what we are finding.

Explanation. Finding functions from $\mathbf{N}^* \to \mathbf{N}^*$ is essentially equivalent to discussing questions on sequences of positive integers $\{f(n)\}$. Here we applied the analysis of prime factors, which is transferred from the method of number theory. Similarly, some ideas in functional equations can also be used in these questions.

Example 4. Given a positive integer k, find all functions f: $\mathbf{N}^* \to \mathbf{N}^*$, satisfying that for any m, $n \in \mathbf{N}^*$,

$$(f(m) + f(n)) \mid (m+n)^k. \qquad ①$$

Solution. Obviously, function $f(n) = n$ satisfies the condition. Then is it the unique one? We are going to prove this.

First, let's prove that f is an injective mapping.

In fact, if there exist a, $b \in \mathbf{N}^*$ such that $a \neq b$ and $f(a) = f(b)$, then it can be inferred from ① that for any $n \in \mathbf{N}^*$,

$$(f(a) + f(n)) \mid (a+n)^k, \ (f(b) + f(n)) \mid (b+n)^k.$$

Hence for any $n \in \mathbf{N}^*$, $f(a) + f(n)$ is the common divisor of $(a+n)^k$ and $(b+n)^k$.

Now take a prime number $p > \max\{a, |b-a|\}$. Let $n = p - a$.

Since

$$(a+n, b+n) = (a+n, b-a) = (p, b-a) = 1,$$

We have $((a+n)^k, (b+n)^k) = 1$. Consequently, $f(a)+f(n) = 1$, which is contrary to the fact that $f(a)$, $f(n)$ are both positive integers. Therefore f is an injective mapping.

Next, for any $m \in \mathbf{N}^*$, let's prove that $|f(m+1)-f(m)| = 1$.

Applying the result of (m, n) and $(m+1, n)$ to ①, we have

$$(f(m)+f(n)) \mid (m+n)^k,$$
$$(f(m+1)+f(n)) \mid (m+1+n)^k.$$

As $(m+n, m+1+n) = 1$, it follows that

$$(f(m)+f(n), f(m+1)+f(n)) = 1.$$

Moreover when m is fixed, for any $n \in \mathbf{N}^*$,

$$(f(n)+f(m), f(m+1)-f(m)) = 1. \quad ②$$

If $|f(m+1)-f(m)| \neq 1$, then there exists prime number p such that $p \mid |f(m+1)-f(m)|$. Now take $\alpha \in \mathbf{N}^*$ such that $n = p^\alpha - m$ is a positive integer. Then it can be inferred from ① that $f(n)+f(m) \mid p^{\alpha k}$ and thus, $f(n)+f(m) = p^l$, where l is a positive integer. Consequently,

$$(f(n)+f(m), f(m+1)-f(m)) = (p^l, f(m+1)-f(m)) \geqslant p.$$

This is contrary to ②.

At last, for any $n \in \mathbf{N}^*$, let's prove that $f(n) = n$.

We can deduce from the previous result that for any $m \in \mathbf{N}^*$, there holds $f(m+1)-f(m) = 1$ or $f(m+1)-f(m) = -1$. If these two situations occur in the same function f satisfying the conditions, then there exists $m \in \mathbf{N}^*$ such that $(f(m+1)-f(m), f(m+2)-f(m+1)) = (1, -1)$ or $(-1, 1)$. Either of the two results leads to the equation that $f(m+2) = f(m)$, which is contrary to that f is an infective mapping. Therefore either of the formula that $f(m+1)-f(m) = 1$ or $f(m+1)-f(m) = -1$ holds for any $m \in \mathbf{N}^*$.

Since f is a function from $\mathbf{N}^* \to \mathbf{N}^*$, it can only hold that for any $m \in \mathbf{N}^*$, $f(m+1) - f(m) = 1$. Therefore for any $n \in \mathbf{N}^*$, $f(n) = n + c$ (where $c = f(1) - 1 \geqslant 0$).

If $c > 0$, we can take a prime number $p > 2c$. By ①, we have $f(1) + f(p-1) \mid p^k$. Then $p + 2c \mid p^k$, which needs $p + 2c$ to be a power of p. Thus, $p \mid p + 2c$. Then $p \mid 2c$, which is contradictory. Hence $c = 0$.

In conclusion, only the function $f(n) = n$ meets the condition.

11 Choosing a Proper Span

From this section on, in the following four sections, we will introduce some common techniques that may be used in proving questions by mathematical induction.

Logic Structure Let $P(n)$ be a proposition (or property) about (of) positive integer n, k a given positive integer. Suppose that

(1) $P(1)$, $P(2)$, $\cdots$, $P(k)$ are true;

(2) it can be inferred from the validity of $P(n)$ that $P(n+k)$ is true.

Then for any $n \in \mathbf{N}^*$, P(n) is true.

Here k is a span. When $k = 1$, this is the first form of mathematical induction. For some questions, it is more convenient to choose a larger span.

Example 1. For any positive integer $n \geqslant 3$, prove that there exists a perfect cube, that can be written as the sum of the cube of n positive numbers.

Proof. We can have an understanding of the background of the question by comparing it with the result that the indeterminate equation $x^3 + y^3 = z^3$ has no positive integer solution.

When $n = 3$, from equality $3^3 + 4^3 + 5^3 = 6^3$, we can deduce that the proposition is true for $n = 3$;

When $n = 4$, from equality $5^3 + 7^3 + 9^3 + 10^3 = 13^3$ (which was discovered first by Euler), we can deduce that the proposition is true for $n = 4$.

Now suppose that the proposition is true for $n (\geqslant 3)$, i.e., there exist positive integers $x_1 < x_2 < \cdots < x_n < y$, satisfying that

$$x_1^3 + x_2^3 + \cdots + x_n^3 = y^3.$$

Then by equality $6^3 = 3^3 + 4^3 + 5^3$, we have

$$\begin{aligned}(6y)^3 &= (6x_n)^3 + \cdots + (6x_2)^3 + (6x_1)^3 \\ &= (6x_n)^3 + \cdots + (6x_2)^3 + (5x_1)^3 + (4x_1)^3 + (3x_1)^3.\end{aligned}$$

This shows that the proposition is true for $n + 2$.

Therefore for any $n \geqslant 3$, the proposition is valid.

Example 2. Let n be a positive integer no less than 3. Prove that an equilateral triangle can be divided into n isosceles triangles.

Proof. When $n = 3$, let O be the excentre of equilateral triangle ABC. Then $\triangle AOB$, $\triangle BOC$ and $\triangle COA$ are all isosceles triangles. Hence the proposition is true for $n = 3$.

When $n = 4$, let D, E, F be the middle point of BC, CA, AB. Then $\triangle AEF$, $\triangle FBD$, $\triangle DCE$, and $\triangle DEF$ are all isosceles triangles. Hence the proposition is true for $n = 4$.

When $n = 5$, as shown in Figure 6, let O be the excentre of equilateral triangle ABC, and D, E the middle point of BC, CA, respectively, and F the middle point of BO. Since the bisector is half of the hypotenuse in a right triangle, $\triangle ABO$, $\triangle BFD$, $\triangle FOD$, $\triangle DEC$, and $\triangle ADE$ are all isosceles triangles. Hence the proposition is true for $n = 5$.

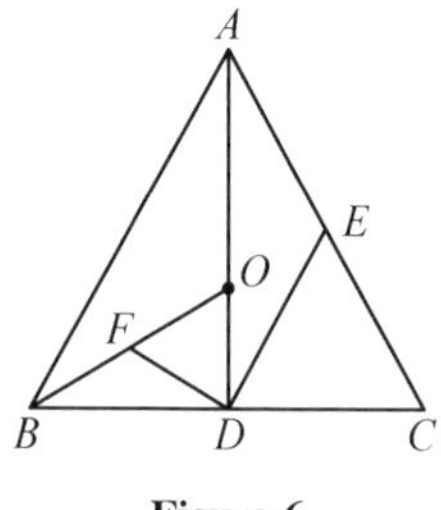

Figure 6

Now suppose that any equilateral triangle can be divided into $n (\geqslant 3)$ isosceles triangles (i.e., the proposition is true for n). Then for equilateral triangle ABC, let D, E, F be the middle point of BC, CA, AB respectively. We can divide equilateral triangle AEF into n

isosceles triangles. Then combining these n triangles with $\triangle BDF$, $\triangle CDE$ and $\triangle DEF$, we divide equilateral triangle ABC into $n+3$ isosceles triangles. Hence the proposition is true for $n+3$.

Generalizing all above, the proposition is true for any $n \geqslant 3$.

Example 3. For any $n \in \mathbf{N}^*$, prove that there are infinitely many positive integer solutions to the indeterminate equation

$$x^2 + y^2 = z^n. \quad ①$$

Proof. When $n=1$, for any $x, y \in \mathbf{N}^*$, (x, y, x^2+y^2) is a positive integer solution to ①; when $n=2$, taking $m > n \geqslant 1$, $m, n \in \mathbf{N}^*$ and letting $x = m^2 - n^2$, $y = 2mn$, $z = m^2 + n^2$, we have $x^2 + y^2 = z^2$. Therefore the proposition is true for $n = 1, 2$.

Now suppose that the proposition is true for n. For positive integers x, y, z, if $x^2 + y^2 = z^n$, then $(xz)^2 + (yz)^2 = z^{n+2}$. Hence there are infinitely many positive integer solutions to the indeterminate equation $x^2 + y^2 = z^{n+2}$. Noting that the proposition is true for $n = 1$, 2, we have that the proposition is true for any $n \in \mathbf{N}^*$.

Explanation. This question can also be solved this way: let $z = a + bi$, where $a, b \in \mathbf{N}^*$ and $0 < \arg z < \frac{\pi}{n}$ (there are infinitely many pairs of a, b such that the values of $a^2 + b^2$ are distinct). Then by the Binomial Theorem, we have $z^n = (a+bi)^n = x + yi$, $x, y \in \mathbf{Z}$ and $xy \neq 0$ (since $\arg z^n \in (0, \pi)$). Taking the modulus of both sides, we get $\left(\sqrt{a^2+b^2}\right)^n = \sqrt{x^2+y^2}$, i.e. $x^2 + y^2 = (a^2+b^2)^n$. Hence $(|x|, |y|, a^2+b^2)$ is a positive integer solution to $x^2 + y^2 = z^n$.

Example 4. Find all of the functions $f: \mathbf{N} \to \mathbf{N}$ satisfying that

(1) For any $m, n \in \mathbf{N}$, there holds that

$$f(m^2 + n^2) = f(m)^2 + f(n)^2;$$

(2) $f(1) > 0$.

Solution. Letting $m = n = 0$ in (1), we have $f(0) = 2f(0)^2$.

Then $f(0) = 0$ or $\frac{1}{2}$. Since $f(0) \in \mathbf{N}$, $f(0) = 0$. Hence, by (1), we have that $f(m^2) = f(m)^2$ holds for any $m \in \mathbf{N}$.

First let's find the value of $f(n)$ for $n \in \{1, 2, \cdots, 10\}$.

From the condition and the previous result, we can deduce that $f(1) = f(1^2) = f(1)^2$. Since $f(1) > 0$, $f(1) = 1$.

Moreover,

$$\begin{aligned} f(2) &= f(1^2 + 1^2) = f(1)^2 + f(1)^2 = 1 + 1 = 2; \\ f(4) &= f(2^2) = f(2)^2 = 4; \\ f(5) &= f(2^2 + 1^2) = f(2)^2 + f(1)^2 = 5; \\ f(8) &= f(2^2 + 2^2) = f(2)^2 + f(2)^2 = 8. \end{aligned}$$

Noting that

$$\begin{aligned} 25 &= f(5)^2 = f(5^2) = f(3^2 + 4^2) \\ &= f(3)^2 + f(4)^2 = f(3)^2 + 16, \end{aligned}$$

and that $f(3) \in \mathbf{N}$, we have $f(3) = 3$. Therefore,

$$\begin{aligned} f(9) &= f(3)^2 = 9, \\ f(10) &= f(3^2 + 1^2) = f(3)^2 + f(1)^2 = 10. \end{aligned}$$

We can find that $f(7) = 7$ with the help of condition (1) and the fact that $7^2 + 1^2 = 5^2 + 5^2$. Then by the fact that $10^2 = 6^2 + 8^2$, we have $f(10)^2 = f(6)^2 + f(8)^2$ and $f(6) = 6$.

Hence, for any $0 \leqslant n \leqslant 10$, $f(n) = n$.

Now let's prove by induction with step length of 5 that for any $n \in \mathbf{N}$, $f(n) = n$.

For this purpose, we need the identities below.

$$\begin{aligned} (5k+1)^2 + 2^2 &= (4k+2)^2 + (3k-1)^2; \\ (5k+2)^2 + 1^2 &= (4k+1)^2 + (3k+2)^2; \\ (5k+3)^2 + 1^2 &= (4k+3)^2 + (3k+1)^2; \\ (5k+4)^2 + 2^2 &= (4k+2)^2 + (3k+4)^2; \\ (5k+5)^2 &= (4k+4)^2 + (3k+3)^2. \end{aligned}$$

Each term on the right side of the identities is less than the first term on the left for $k \geqslant 2$. Thus, by condition (1) and the inductive hypothesis, we can determine the function values of the first terms on the left side of the identities. That is, the proposition is true as we make use of induction every five numbers.

Therefore, for any $n \in \mathbf{N}$, $f(n) = n$.

Explanation. From the example above, we can find that proving that $P(n)$ is true by induction with step length of k is essentially partitioning $\{P(n)\}$ into k sets of propositions and proving them respectively. It is certain that we can combine this idea with the second form of mathematical induction and make use of one set of propositions to prove another. This idea is shown in the example above.

12 Choosing the Appropriate Object for Induction

Statements that are relevant to positive integers will sometimes involve multiple variables. When we deal with them using mathematical induction, we need to decide which of the objects is the right one to conduct mathematical induction on first.

Example 1. Suppose that $m, n \in \mathbf{N}^*$. Prove that for any positive real numbers $x_1, \cdots, x_n; y_1, \cdots, y_n$, if

$$x_i + y_i = 1, \; i = 1, 2, \cdots, n,$$

then $(1 - x_1 \cdots x_n)^m + (1 - y_1^m)(1 - y_2^m) \cdots (1 - y_n^m) \geqslant 1.$ ①

Proof. We make use of mathematical induction on n. When $n = 1$, from the given condition, we know that

$$(1 - x_1)^m + (1 - y_1^m) = y_1^m + (1 - y_1^m) = 1.$$

So ① holds true for $n = 1$.

Now we suppose ① is true for some $n - 1 (n \geqslant 2)$. We consider n.

$$\begin{aligned}
&(1 - x_1 \cdots x_n)^m + (1 - y_1^m) \cdots (1 - y_n^m) \\
=&(1 - x_1 \cdots x_{n-1}(1 - y_n))^m + (1 - y_1^m) \cdots (1 - y_n^m) \\
\geqslant&(1 - x_1 \cdots x_{n-1} + x_1 \cdots x_{n-1} y_n)^m + (1 - (1 - x_1 \cdots x_{n-1})^m)(1 - y_n^m).
\end{aligned}$$

Let us denote $a = 1 - x_1 \cdots x_{n-1}$, $b = y_n$. From the above equations, we know that in order to prove ① to be true for n, we only need to prove:

$$(a + b - ab)^m + (1 - a^m)(1 - b^m) \geqslant 1$$

holds true for any a, $b \in (0, 1)$. That is to prove

$$(a + b - ab)^m \geqslant a^m + b^m - a^m b^m. \quad ②$$

To deal with ②, we proceed by inducting on m.

When $m = 1$, obviously ② is true. Now we suppose ② to be true for some $m - 1 (m \geqslant 2)$. Then,

$$\begin{aligned}
&(a + b - ab)^m - a^m - b^m + a^m b^m \\
\geqslant&(a^{m-1} + b^{m-1} - a^{m-1}b^{m-1})(a + b - ab) - a^m - b^m + a^m b^m \\
=&2a^m b^m + ab^{m-1} + ba^{m-1} - a^m b^{m-1} - a^{m-1}b^m - a^m b - ab^m \\
=&(b^{m-1} - b^m)(a - a^m) + (a^{m-1} - a^m)(b - b^m).
\end{aligned}$$

We notice that a, $b \in (0, 1)$, so

$$b^{m-1} \geqslant b^m,\ a \geqslant a^m,\ a^{m-1} \geqslant a^m,\ b \geqslant b^m.$$

Therefore,

$$(a + b - ab)^m \geqslant a^m + b^m - a^m b^m,$$

i.e., ② holds true for m.

From this, we know that ① is true.

Explanation. This is a problem involving two variables that are both positive integers. It naturally comes to mind that we induct with respect to n while treating m as a constant, because the second addend on the left side of ① seems easier to deal with while in the process of inducting on n.

Example 2. Prove that for any m, $n \in \mathbf{N}^*$,

$$S(m, n) = \sum_{i=0}^{2^{n-1}-1} \left(\tan \frac{(2i + 1)\pi}{2^{n+1}}\right)^{2m}$$

is a positive integer.

Proof. We choose n to be our object of induction.

When $n=1$, $S(m, 1)=\left(\tan \frac{\pi}{4}\right)^{2m}=1$, so the statement is true for $n=1$ for all $m \in \mathbf{N}^*$.

Now we suppose the statement is true for some $n-1$ for all $m \in \mathbf{N}^*$. We consider the case for n.

From $\cot 2\alpha=\frac{1-\tan^2\alpha}{2\tan\alpha}=\frac{1}{2}(\cot\alpha-\tan\alpha)$, we know that

$$\tan^2\left(\frac{\pi}{2}-2\alpha\right)=\frac{1}{4}(\cot\alpha-\tan\alpha)^2=\frac{1}{4}(\tan^2\alpha+\cot^2\alpha-2).$$

So, we rewrite the summation as a pairing of the beginning and ending terms of the original sum (i. e. the paired sum of the ith and $(2^{n-1}-1-i)^{\text{th}}$ terms) to see that:

$$\begin{aligned}
S(1, n) &= \frac{1}{2}\sum_{i=0}^{2^{n-1}-1}\left(\tan^2\frac{(2i+1)\pi}{2^{n+1}}+\tan^2\frac{(2(2^{n-1}-1-i)+1)\pi}{2^{n+1}}\right)\\
&= \frac{1}{2}\sum_{i=0}^{2^{n-1}-1}\left(\tan^2\frac{(2i+1)\pi}{2^{n+1}}+\tan^2\left(\frac{\pi}{2}-\frac{(2i+1)\pi}{2^{n+1}}\right)\right)\\
&= \frac{1}{2}\sum_{i=0}^{2^{n-1}-1}\left(\tan^2\frac{(2i+1)\pi}{2^{n+1}}+\cot^2\frac{(2i+1)\pi}{2^{n+1}}\right)\\
&= \frac{1}{2}\sum_{i=0}^{2^{n-1}-1}\left(4\tan^2\frac{(2i+1)\pi}{2^{n}}+2\right)\\
&= 2\sum_{i=0}^{2^{n-1}-1}\tan^2\frac{(2i+1)\pi}{2^{n}}+2^{n-1}\\
&= 2\sum_{i=0}^{2^{n-2}-1}\left(\tan^2\frac{(2i+1)\pi}{2^{n}}+\tan^2\frac{(2(2^{n-1}-1-i)+1)\pi}{2^{n}}\right)+2^{n-1}\\
&= 2\sum_{i=0}^{2^{n-2}-1}\left(\tan^2\frac{(2i+1)\pi}{2^{n}}+\tan^2\left(\pi-\frac{(2i+1)\pi}{2^{n}}\right)\right)+2^{n-1}\\
&= 4\sum_{i=0}^{2^{n-2}-1}\tan^2\frac{(2i+1)\pi}{2^{n}}+2^{n-1}=4S(1, n-1)+2^{n}.
\end{aligned}$$

So $S(1, n) \in \mathbf{N}^*$.

Next, we suppose $S(1, n)$, $S(2, n)$, $\cdots$, $S(m-1, n)$ are all positive integers. Thus, we consider $S(m, n)$ noting that for all $k \in \mathbf{N}$, we have

$$(x + x^{-1})^k = C_k^0(x^k + x^{-k}) + C_k^1(x^{k-1} + x^{-(k-1)}) + \cdots, \qquad ①$$

by the Binomial Theorem.

Therefore,

$$\left(\frac{1}{4}(x + x^{-1} - 2)\right)^m$$

$$= \frac{1}{4^m}\sum_{k=0}^{m} C_m^k (x + x^{-1})^k \cdot (-2)^{m-k}$$

$$= \frac{1}{4^m}((x^m + x^{-m}) + b_1(x^{m-1} + x^{-(m-1)}) + \cdots + b_{m-1}(x + x^{-1}) + b_m),$$

as from ①, where $b_1, \cdots, b_m \in \mathbf{Z}$.

Let $x = \tan^2 \frac{(2i+1)\pi}{2^{n+1}}$ for the above equation, and we sum the terms up for $i = 0, 1, 2, \cdots, 2^{n-2} - 1$. By making use of calculations similar to $S(1, n)$ we may know that

$$S(m, n-1) = \frac{1}{4^m}(S(m, n) + b_1 S(m-1, n) + \cdots + b_{m-1}S(1, n) + b_m),$$

then we have

$$S(m, n) = 4^m \cdot S(m, n-1) - b_1 S(m-1, n) - \cdots - b_{m-1}S(1, n) - b_m.$$

Therefore, $S(m, n) \in \mathbf{Z}$. Since every term of $S(m, n)$ is greater than zero, we have $S(m, n) \in \mathbf{N}^*$.

Generalizing from above, for any $m, n \in \mathbf{N}^*$, all $S(m, n)$ are positive integers. The statement is thus proved.

Explanation. Essentially, we adopted a method of inducting with respect to m, a second variable, during the process of inducting from $n-1$ to n. In statements where two variables are involved, the method we have used is one of the common ways that we follow when using the principles of mathematical induction.

Example 3. Suppose the non-negative integers $a_1, a_2, \cdots, a_t$

satisfy

$$a_i + a_j \leqslant a_{i+j} \leqslant a_i + a_j + 1,$$

where $1 \leqslant i, j \leqslant t, i + j \leqslant t$.

Prove that there exists $x \in \mathbf{R}$ such that for all $n \in \{1, 2, \cdots, t\}$, we have $a_n = [nx]$.

Proof. Denote $I_n = \left[\frac{a_n}{n}, \frac{a_n + 1}{n}\right)$, for $n = 1, 2, \cdots, t$. We need to prove that there exists a real number $x \in \bigcap_{n=1}^{t} I_n$. ①

Let $L = \max_{1 \leqslant n \leqslant t} \frac{a_n}{n}$, $U = \min_{1 \leqslant n \leqslant t} \frac{a_n + 1}{n}$. If we are able to prove $L < U$, then ① is true since $\bigcap_{n=1}^{t} I_n = [L, U)$. In the meantime, in order to prove $L < U$ true, we only need to prove the following: for any m, $n \in \{1, 2, \cdots, t\}$, we have $\frac{a_n}{n} < \frac{a_m + 1}{m}$, namely,

$$ma_n < n(a_m + 1). \qquad ②$$

Next, we prove ② true by inducting on $m + n$.

When $n + m = 2$, we have $m = n = 1$. Then ② is obviously true. Suppose ② holds for all positive integer pairs (m, n) which satisfy $n + m \leqslant k$. Then when $n + m = k + 1$, if $m = n$, then ② is obviously true. If $m > n$, then we know from the induction hypothesis that $(m - n)a_n < n(a_{m-n} + 1)$. We know $n(a_{m-n} + a_n) \leqslant na_m$ from $a_i + a_j \leqslant a_{i+j}$ by the given conditions. Therefore, $ma_n < n(a_m + 1)$, i.e., ② is true. If $m < n$, then we know $ma_{n-m} < (n - m)(a_m + 1)$ by the induction hypothesis. We may deduce that $ma_n \leqslant m(a_m + a_{n-m} + 1)$ from $a_{i+j} \leqslant a_i + a_j + 1$ given by the conditions. So $ma_n < n(a_m + 1)$, i.e., ② holds true.

Generalizing all above, for any m, $n \in \{1, 2, \cdots, t\}$, ② is always true.

Explanation. Generally speaking, for any $x \in \mathbf{R}$, $i, j \in \mathbf{N}^*$, it is always true that $[ix] + [jx] \leqslant [(i + j)x] \leqslant [ix] + [jx] + 1$. This actually is a property of the greatest integer function, and we come to the conclusion of this problem when we discuss it the other way around. Please be aware that the conclusion from this problem is only valid for an arbitrary number of finite terms and we are not able to find an x which satisfies the requirement for an infinite sequence a_1, a_2, $\cdots$ with properties described by the problem.

From a point of view regarding methodology, the strategy of inducting with respect $m+n$ is convenient and appropriate.

Example 4. Suppose m, n are distinct positive integers. A sequence comprised of integers satisfy the following condition: the sum of any consecutive m terms is negative, while the sum of any consecutive n terms is positive. Please state at most how many terms this sequence has.

Solution. Suppose $(m, n)=d$, $m=m_1 d$, $n=n_1 d$, then $(m_1, n_1)=1$. We also suppose that the sequence a_1, $\cdots$, a_t satisfies all conditions. Denote $A_i=a_{(i-1)d+1}+\cdots+a_{id}$, $i=1, 2, \cdots$.

On one hand, if $t \geqslant (m_1+n_1-1)d$, we consider the following number array

$$\begin{array}{l} A_1, A_2, \cdots, A_{m_1};\\ A_2, A_3, \cdots, A_{m_1+1};\\ \cdots\cdots\cdots\cdots\cdots\cdots\cdots\cdots\\ A_{n_1}, A_{n_1+1}, \cdots, A_{n_1+m_1-1}. \end{array}$$

From the given conditions, the sum of all numbers on each row is negative, and the sum of each column is positive. Note that the sum of all the numbers in this array should be negative if by row but the sum is positive if by column. This is self-contradictory. Thus, we may deduce that $t \leqslant (m_1+n_1-1)d-1$, i.e., $t \leqslant m+n-(m, n)-1$.

On the other hand, we need to prove: there exists an integer sequence with length of $m+n-(m, n)-1$ that meets the requirements. For that, we need to prove the statement below:

Proposition. Suppose $d \in \mathbf{N}^*$, and s, t are distinct positive integers such that $(s, t)=1$, then there exists a rational number sequence of length $(s+t-1)d-1$, the sum of whose arbitrary consecutive sd terms is a negative number, and the sum of whose arbitrary td terms is a positive number. (Attention: If we multiply every number of this sequence in this proposition by the common denominator of all numbers in sequence, then we can get a desired integer sequence.)

We use mathematical induction in terms of $\max\{s, t\}=r$.

When $\max\{s, t\}=2$, without loss of generality let $s=2$, $t=1$ (otherwise, we may multiply every number in the sequence by -1), then we may take arbitrarily $2d-1$ positive rational numbers so that we may get a sequence which satisfies the proposition.

Suppose that the proposition is correct for $\max\{s, t\}<r$ $(r\geqslant 3)$, then we consider the case $\max\{s, t\}=r$.

Without loss of generality, we may let $s>t$, and notice that $(s-t, t)=1$. Then, by the induction hypothesis, there exists a rational number sequence $b_1, b_2, \cdots, b_{(s-1)d-1}$, whose sum of arbitrary consecutive $(s-t)d$ terms is a negative number, and the sum of whose consecutive td terms is a negative number. We prove that there exist rational numbers $a_1, a_2, \cdots, a_{td}$ such that the system of inequalities below holds:

$$\begin{cases} a_{d+1}+\cdots+a_1+b_1+\cdots+b_{(s-1)d-1}<0, \\ a_{d+2}+\cdots+a_1+b_1+\cdots+b_{(s-1)d-2}<0, \\ \cdots\cdots\cdots\cdots\cdots\cdots\cdots\cdots\cdots\cdots \\ a_{td}+\cdots+a_1+b_1+\cdots+b_{(s-t)d}<0. \end{cases} \quad ①$$

and,

$$\begin{cases} a_{td}+\cdots+a_1>0, \\ a_{td-1}+\cdots+a_1+b_1>0, \\ \cdots\cdots\cdots\cdots\cdots\cdots\cdots \\ a_1+b_1+\cdots+b_{td-1}>0. \end{cases} \quad ②$$

Then, the sequence $a_{td}, a_{td-1}, \cdots, a_1, b_1, \cdots, b_{(s-1)d-1}$ is a sequence that satisfies the proposition with length $(s+t-1)d-1$. So, the proposition is proved.

As a matter of fact, if the systems of inequalities are to hold simultaneously, we only need to choose rational numbers a_1, a_2, $\cdots$, a_d, such that

$$a_1 > -(b_1 + \cdots + b_{td-1}),\ a_2 > -(a_1 + b_1 + \cdots + b_{td-2}),\ \cdots,$$
$$a_d > -(a_{d-1} + \cdots + a_1 + b_1 + \cdots + b_{(t-1)d}),$$

then we take a_{d+1} such that it is a rational number satisfying

$$-(a_d + \cdots + a_1 + b_1 + \cdots + b_{(t-1)d-1}) < a_{d+1} < -(a_d + \cdots + a_1 + b_1 + \cdots + b_{(s-1)d-1}). \quad ③$$

Note that the right-hand side of ③ when subtracted from it the left-hand side will equal $-(b_{(t-1)d} + \cdots + b_{(s-1)d-1}) > 0$ (here we used the induction hypothesis). Therefore, an a_{d+1} satisfying the conditions does exist. Then we can deduce in the same manner rational numbers a_1, $\cdots$, a_{td} satisfying the systems of inequalities ① and ② exist.

We go back to the original question. The integer sequence that satisfies the conditions has at most $m + n - (m, n) - 1$ terms.

Explanation. The cases $m = 11$, $n = 6$ for this question once appeared as a competition question. While the example for $m = 11$, $n = 6$ is easy to get, it is hard to find a generic example for m, n. When this example appeared on a quiz in the China National Math Olympiad Trainee Team in the year 2000, a very small number of students correctly solved it.

13 Make Appropriate Changes to the Propositions

When we use mathematical induction to prove propositions, we sometimes need to deal with them by strengthening conditions, making use of auxiliary propositions, or making the proposition more general, etc.

Example 1. For any positive integer n, we have

$$\frac{1}{2}\cdot\frac{3}{4}\cdot\cdots\cdot\frac{2n-1}{2n}<\frac{1}{\sqrt{3n}}. \quad ①$$

Proof. If we deal with it directly using weak induction, during the inductive step we would need to show that the inequality

$$\frac{2n+1}{2(n+1)}\cdot\frac{1}{\sqrt{3n}}\leqslant\frac{1}{\sqrt{3(n+1)}}$$

is true, which requires $(n+1)(2n+1)^2 \leqslant n(2n+2)^2$, and this is equivalent to $(2n+1)^2 \leqslant n(4n+3)$. However, this inequality is not true. So it is hard to prove ① to be true by directly applying mathematical induction.

We instead prove a strengthened proposition of ①:

$$\frac{1}{2}\cdot\frac{3}{4}\cdot\cdots\cdot\frac{2n-1}{2n}\leqslant\frac{1}{\sqrt{3n+1}}. \quad ②$$

When $n=1$, the left-hand-side of ② is equal to $\frac{1}{2}$, and the right-hand side is equal to $\frac{1}{2}$. So ② is true for $n=1$.

Now we suppose ② is true for some n, then for $n+1$, we have

$$\frac{1}{2}\cdot\frac{3}{4}\cdot\cdots\cdot\frac{2n-1}{2n}\cdot\frac{2n+1}{2(n+1)}\leqslant\frac{1}{\sqrt{3n+1}}\cdot\frac{2n+1}{2(n+1)}.$$

In order to prove ② is true for $n+1$, we only need to prove that

$$\frac{1}{\sqrt{3n+1}}\cdot\frac{2n+1}{2n+2}\leqslant\frac{1}{\sqrt{3n+4}}.$$

That is to prove

$$(3n+4)(2n+1)^2\leqslant(3n+1)(2n+2)^2. \quad ③$$

We note that ③ is equivalent to

$$\begin{aligned}3(2n+1)^2 &\leqslant (3n+1)((2n+2)^2-(2n+1)^2)=(3n+1)(4n+3)\\ &\Leftrightarrow 12n^2+12n+3\leqslant 12n^2+13n+3\\ &\Leftrightarrow n\geqslant 0.\end{aligned}$$

So, ③ is true. Therefore ② is true for $n+1$, that is to say, for an arbitrary $n \in \mathbf{N}^*$, we have ② holds true.

Considering $\sqrt{3n+1} > \sqrt{3n}$, we know that ① is true for any $n \in \mathbf{N}^*$.

Explanation. It is sometimes difficult to realize the inductive step from n to $n+1$ when using mathematical induction for proposition $P(n)$, where n is a positive integer. However, it is sometimes easier for a strengthened proposition $Q(n)$ to be proved using mathematical induction. Therefore, we sometimes need to strengthen the proposition ourselves. Of course, when we do that, we need to make appropriate choices under the premise of getting the essence of the proposition. The aim is to help realize the inductive step for induction.

Example 2. Suppose $A_1, A_2, \cdots, A_r$ is an arbitrary r-partition for $\mathbf{N}^*$ (i.e., intersection of any two from $A_1, \cdots, A_r$ is an empty set and $\bigcup_{i=1}^{r} A_i = \mathbf{N}^*$.) Prove: There is a set A among $A_1, \cdots, A_r$ which has the following property: There exists an $m \in \mathbf{N}^*$ such that for any $k \in \mathbf{N}^*$ such that we can take k numbers $a_1, \cdots, a_k$ from A, satisfying that for $1 \leqslant j \leqslant k-1$, we have $1 \leqslant a_{j+1} - a_j \leqslant m$.

Proof. Suppose that $P \subseteq \mathbf{N}^*$. If there are connected segments of positive integers of arbitrary lengths in P, then we call P a Long Subset.

We will strengthen the proposition as follows:

For any Long Subset P, any r-partition $A_1, A_2, \cdots, A_r$ on P must have one set A which enjoys the property required of the problem.

We apply mathematical induction with respect to r.

When $r = 1$, according to the definition of Long Subset, we know that the proposition is correct by taking $m = 1$.

We suppose the proposition is true for the case $r = n$. We consider the case $r = n+1$.

Suppose $P = (A_1 \cup A_2 \cup \cdots \cup A_n) \cup A_{n+1}$, $Q = A_1 \cup A_2 \cup \cdots \cup A_n$. If Q is a Long Subset, we know that the proposition is correct by the induction hypothesis. If Q is not a Long Subset, then there must exist an $l \in \mathbf{N}^*$ such there is no consecutive positive integer segments of length l in Q. Since P is a Long Subset, for any $k \in \mathbf{N}^*$, there exists a consecutive positive integer segment of length kl, within which there are at least k numbers which belong to A_{n+1}. Now we take out the least k numbers belonging to A_{n+1} from this consecutive positive integer segment of length kl. Then the difference of two consecutive numbers in P is no bigger than $2l$. Then, we take $m = 2l$, and the set A_{n+1} enjoys the property required of the problem.

For all above, the strengthened proposition is proved true. Since $\mathbf{N}^*$ itself is a Long Subset, the original proposition is correct.

Explanation. The problem essentially requires to prove that for each r - Partition of $\mathbf{N}^*$, there exist sets A and $m \in \mathbf{N}^*$, such that after we partition the numbers in $\mathbf{N}^*$ into continuous integer segments of length $\frac{m}{2}$, for any $k \in \mathbf{N}^*$, there are adjacent k "consecutive integer segments" such that each of the "consecutive integer segments" has within it a number that belongs to A. Hence, if the union set of other subsets does not contain consecutive integer segments of arbitrary lengths, then we can find k numbers that satisfy the given conditions within A. By this we thought of introducing the concept of "Long Subset", and then appropriately strengthened the problem.

Example 3. Prove that: there exist infinitely many $n \in \mathbf{N}^*$ such that

$$n \mid (2^n + 2). \qquad ①$$

Proof. $n = 2$ satisfies ①. The next integer that satisfies ① is positive integer $n = 6$. The relationship between them is $6 = 2^2 + 2$. This hints us to deal with this problem using the following method.

Suppose $n(>1)$ is a positive integer with property ①, if we are able to prove: $(2^n + 2) \mid (2^{2^n+2} + 2)$, then we may deduce in this manner that there are infinitely many positive integers n that satisfy ①.

We note that $(2^{n-1}+1) \mid (2^{2^n+1}+1)$ holds under the condition $(n-1) \mid (2^n+1)$. We deal with it by adding one more requirement.

Noting that $n=2$ enjoys the property, now we suppose $n(\geqslant 2)$ enjoys the above property. Let $m=2^n+2$, we are to prove m also enjoys the above property.

As a matter of fact, since $(n-1) \mid (2^n+1)$, and 2^n+1 is an odd number, we may suppose $2^n+1=(n-1)q$, where q is an odd number. Then

$$\begin{aligned} 2^{m-1}+1 &= 2^{2^n+1}+1=(2^{n-1})^q+1 \\ &=(2^{n-1}+1)((2^{n-1})^{q-1}-(2^{n-1})^{q-2}+\cdots+1), \end{aligned}$$

so $(2^{n-1}+1) \mid (2^{m-1}+1)$, and also $(2^n+2) \mid (2^m+2)$, i.e., $m \mid (2^m+2)$.

On the other hand, by $(n-1) \mid (2^n+1)$, we know that $n-1$ is an odd number so n is an even number. Thus, by $n \mid (2^n+2)$, we may assume that $2^n+2=np$, where p is an odd number (here we use $4 \nmid (2^n+2)$), then,

$$2^m+1=(2^n)^p+1=(2^n+1)((2^n)^{p-1}-(2^n)^{p-2}+\cdots+1),$$

i.e., we have $(2^n+1) \mid (2^m+1)$, and that leads to $(m-1) \mid (2^m+1)$.

Generalizing above, we know that the proposition is true.

Further thought:

If the problem is like follows: Prove that there exist infinitely many positive integers $n(>1)$ such that $n-1 \mid 2^n+1$. Do we still need to strengthen it to be proposition ②?

Example 4. Find all the functions $f: \mathbf{Z} \to \mathbf{Z}$ such that for any x, y, $z \in \mathbf{Z}$, we have

$$f(x^3+y^3+z^3)=f(x)^3+f(y)^3+f(z)^3.$$

Solution. It's not hard to see that the following three functions

$$f(x)=0, \ f(x)=x, \ f(x)=-x$$

satisfy the conditions in the question.

Next, we are to prove that they are exhaustive.

Take $(x, y, z) = (0, 0, 0)$, we get $f(0) = 3f(0)^3$. This equation of the third degree about $f(0)$ has only one integer solution. So $f(0) = 0$. Then we take $(x, y, z) = (x, -x, 0)$ and we may get $f(x) = -f(-x)$, so $f(x)$ is an odd function. Meanwhile, we let $(x, y, z) = (1, 0, 0)$, and we may get $f(1) = f(1)^3$, then $f(1) \in \{-1, 0, 1\}$.

Next, we are to use mathematical induction to prove:

For any $x \in \mathbf{Z}$, we have $f(x) = f(1)x$ (Then by considering the value of $f(1)$ we complete the solution of this problem.)

We conduct induction with respect to $|x|$. Let $(x, y, z) = (1, 1, 0)$, we have $f(2) = 2f(1)^3 = 2f(1)$. Let $(x, y, z) = (1, 1, 1)$ and we have $f(3) = 3f(1)$. Then considering the fact that $f(x)$ is an odd function, we know that the conclusion ① is true for $|x| \leqslant 3$.

Now we suppose that $f(x) = f(1)x$ is true for $|x| < k(k \in \mathbf{N}^*, k > 3)$. We discuss the case for $f(k)$ and $f(-k)$. Since $f(-k)$ is an odd function, we only need to prove $f(k) = f(1)k$.

For this aim, we need to use the auxiliary proposition below.

Proposition. For any $k \in \mathbf{N}^*$, $k \geqslant 4$, the number k^3 can be expressed as the sum of five cubes, and every term of the five addends has its absolute value less than k^3.

As a matter of fact, from

$$\begin{aligned}
4^3 &= 3^3 + 3^3 + 2^3 + 1^3 + 1^3, \\
5^3 &= 4^3 + 4^3 + (-1)^3 + (-1)^3 + (-1)^3, \\
6^3 &= 5^3 + 4^3 + 3^3 + 0^3 + 0^3, \\
7^3 &= 6^3 + 5^3 + 1^3 + 1^3 + 0^3.
\end{aligned}$$

For odd numbers that are no less than 9, namely, $2m + 1(m \in \mathbf{N}^*, m \geqslant 4)$, we have

$$(2m + 1)^3 = (2m - 1)^3 + (m + 4)^3 + (4 - m)^3 + (-5)^3 + (-1)^3. \quad ②$$

So, the proposition is true for $k = 4$ or 6 and odd numbers that are no less than 3.

Note that for any $k > 3$, $k \in \mathbf{N}^*$, there exists the factored form

$k = my$, where $m \in \mathbf{N}^*$, $y = 4$ or 6 or an odd number bigger than 3. By what we proved before, there is an expression $y^3 = y_1^3 + \cdots + y_5^3$, in which $|y_i| < y$, $1 \leqslant i \leqslant 5$. Then $k^3 = (my_1)^3 + \cdots + (my_5)^3$, and $|my_i| < my = k$. So, the auxiliary proposition holds true.

By the above proposition, for any $k > 3$, $k \in \mathbf{N}^*$, we may know $k^3 = x_1^3 + \cdots + x_5^3$, $|x_i| < k$. So we know from conditions that

$$f(k)^3 + f(-x_4)^3 + f(-x_5)^3 = f(x_1)^3 + f(x_2)^3 + f(x_3)^3.$$

Combining the induction hypothesis, $f(x_i) = f(1)x_i$, $f(-x_i) = -f(1)x_i$, we have

$$f(k)^3 = \sum_{i=1}^{5} f(x_i)^3 = f(1)^3 \sum_{i=1}^{5} x_i^3 = k^3 f(1)^3,$$

so $f(k) = f(1)k$.

So the conclusion ① is proved, and the problem is solved.

Explanation. This problem essentially was made from the identity ②. In the process of proof, the method of introducing an auxiliary proposition is sometimes used for the sake of realizing inductive step for induction, but this is not only intended for proving propositions by mathematical induction. A math problem, no matter how hard it is, is often integrated by creatively combining some simple conclusions.

Example 5. There is one black and one white ball in a jar. We also have another 50 white and 50 black balls. We conduct the following operation for 50 times: Randomly pick out a ball and then put in two balls that have the same color as the one picked. Finally, there are 52 balls in the jar. What is the most likely number of white balls in the jar at the end?

Solution. We prove that for any $1 \leqslant k \leqslant 51$, the probability of having k white balls is always $\frac{1}{51}$.

We make the problem more general. After n operations, the probability of having k white balls is $P_n(k)$, $1 \leqslant k \leqslant n+1$.

Next, we are to prove that

$$P_n(1) = P_n(2) = \cdots = P_n(n+1) = \frac{1}{n+1}.$$

When $n = 1$, the above proposition is obviously true. We suppose the proposition to be true for n, then we consider the case $n+1$. We note that the following recursive formula holds.

$$P_{n+1}(k) = \frac{k-1}{n+2}P_n(k-1) + \frac{n+2-k}{n+2}P_n(k),$$

where $1 \leqslant k \leqslant n+1$, and $P_n(0) = 0$. (The recursive formula is obtained by conducting classified discussions for the number of white balls ($k-1$ and k) in the jar before the $n+1^{\text{th}}$ operation.)

Then, through $P_n(1) = P_n(2) = \cdots = P_n(n+1) = \frac{1}{n+1}$ (induction hypothesis) we know that

$$P_{n+1}(1) = P_{n+1}(2) = \cdots = P_{n+1}(n+1) = \frac{n+1}{n+2} \cdot \frac{1}{n+1} = \frac{1}{n+2}.$$

By combining the fact that $\sum_{k=1}^{n+2} P_{n+1}(k) = 1$, we can prove that $P_{n+1}(n+2) = \frac{1}{n+2}$.

So, the proposition holds true.

Explanation. To make the proposition more general is only a means to an end. Here we do it because we want to make use of the recursive method. Idea and connotation determine the form we present.

Example 6. Prove that there exist positive integers $n_1 < n_2 < \cdots < n_{50}$ such that

$$n_1 + S(n_1) = n_2 + S(n_2) = \cdots = n_{50} + S(n_{50}).$$

Here $S(n)$ represents the sum of the digits of the number as represented by the decimal system.

Proof. We make the proposition more general. We use mathematical induction to prove the conclusion below.

For any $k \in \mathbf{N}^*$, $k \geqslant 2$, there exist positive integers $n_1 < n_2 < \cdots <$

n_k such that

$$n_1 + S(n_1) = n_2 + S(n_2) = \cdots = n_k + S(n_k) \equiv 7 \pmod 9). \quad ①$$

When $k = 2$, we take $n_1 = 107$, $n_2 = 98$. Noting that $107 + 8 = 98 + 17 = 115 \equiv 7(\bmod 9)$, we know that the proposition is true for $k = 2$.

We suppose that the proposition is true for $k(\geqslant 2)$, and we also suppose that $n_1 < n_2 < \cdots < n_k$ satisfies ①. We consider the case $k+1$.

Let $m \in \mathbf{N}^*$, and we make $9m - 2 = n_i + S(n_i)$, $1 \leqslant i \leqslant k$. We take positive integer $n'_i = 9 \times 10^m + n_i$, $1 \leqslant i \leqslant k$, $n'_{k+1} = 8\underbrace{9\cdots9}_{m}$, then $n'_i\ (1 \leqslant i \leqslant k+1)$ are all positive integers with $m+1$ digits, (Be careful that, for $k = 2$, it is obvious that for k in the induction hypothesis, we have $n_i < 10^m$, so n'_i is a number with $m+1$ digits, $1 \leqslant i \leqslant k$), and that for $1 \leqslant i \leqslant k$, we always have

$$n'_i + S(n'_i) = 9 \times 10^m + n_i + (9 + S(n_i)) = 9 \times 10^m + 9m + 7;$$

At the same time,

$$n'_{k+1} + S(n'_{k+1}) = (9 \times 10^m - 1) + (8 + 9m) = 9 \times 10^m + 9m + 7.$$

So, $n'_1 + S(n'_1) = \cdots = n'_{k+1} + S(n'_{k+1}) \equiv 7(\bmod 9)$, and we can conclude from induction hypothesis and the made structure that $n'_{k+1} < n'_1 < n'_2 < \cdots < n'_k$. Therefore the proposition is true for $k+1$.

By all above, the proposition holds true.

Explanation. Here in ①, $n_i + S(n_i) \equiv 7(\bmod 9)$ is required because it is important to find n'_{k+1} for the inductive step. It is a necessary strengthening found when in the process of structuring the induction.

14 Guessing Before Proving

Guessing before proving is a basic procedure for math discoveries. If the guessed proposition cannot be proved then it becomes a mathematical conjecture. By first using case study on small examples to examine a proposition in terms of positive integers n, then using methods like analogy and incomplete induction, we may guess a general conclusion, and attempt to prove it through mathematical induction. This kind of process often occurs in our problem-solving experiences.

Example 1. We have a function defined by $f: \mathbf{N}^* \to \mathbf{N}$, and $f(1) = 0$. For any $n \in \mathbf{N}^*$, $n \geqslant 2$, we have

$$f(n) = \max\left\{f(j) + f(n-j) + j \mid j = 1, 2, \cdots, \left[\frac{n}{2}\right]\right\}.$$

Find the value of $f(2004)$ and prove it to be correct.

Solution. Let's try calculating the values of $f(n)$ for small values of n, and we may find the following results: $f(2) = 1$, $f(3) = 2$, $f(4) = 4$, $f(5) = 5$, $\cdots$. During the process of calculating these values, we may find that when $1 \leqslant j \leqslant \left[\frac{n}{2}\right]$, the maximum value of $f(j) + f(n-j) + j$ is reached when $j = \left[\frac{n}{2}\right]$. Therefore, we may guess

$$f(2n) = 2f(n) + n, \; f(2n+1) = f(n) + f(n+1) + n. \quad ①$$

Next, we will prove ① to be true by mathematical induction.

When $n = 1$, we know from the above discussions that ① is true.

Now we suppose that ① is true for $1, 2, \cdots, n-1$. We consider the case n.

We first find the value of $f(2n)$.

$$\begin{aligned} f(2n) &= \max\{f(j) + f(n-j) + j \mid 1 \leqslant j \leqslant n\} \\ &\geqslant f(n) + f(2n-n) + n \\ &= 2f(n) + n. \end{aligned}$$

Hence, we only need to prove that $f(2n) \leqslant 2f(n) + n$.

Now we discuss the cases $1 \leqslant j \leqslant n$, and separate the cases when j is even or odd.

When $j = 2k$, $1 \leqslant k \leqslant \left[\frac{n}{2}\right]$, by induction hypothesis, we have the following:

$$f(j)+f(2n-j)+j = f(2k)+f(2(n-k))+2k$$
$$= (2f(k)+k)+(2f(n-k)+n-k)+2k$$
$$= 2(f(k)+f(n-k)+k)+n \leqslant 2f(n)+n.$$

The final inequality is obtained through the definition of $f(n)$.

When $j = 2k-1$, $1 \leqslant k \leqslant \left[\frac{n+1}{2}\right]$, we know from induction hypothesis that

$$f(j)+f(2n-j)+j$$
$$= f(2k-1)+f(2(n-k)+1)+2k-1$$
$$= (f(k)+f(k-1)+k-1)+(f(n-k)+f(n-k+1)+n-k)+2k-1$$
$$= (f(k-1)+f(n-(k-1))+k-1)+(f(k)+f(n-k)+k)+n-1$$
$$\leqslant f(n)+f(n)+n = 2f(n)+n.$$

Here we deem $f(0)=0$. When n is even, $\left[\frac{n+1}{2}\right]=\left[\frac{n}{2}\right]$; When n is odd, assume $n=2m+1$, then when $k=\left[\frac{n+1}{2}\right]$,

$$f(k)+f(n-k)+k = f(m+1)+f(m)+m+1 \leqslant f(n)+1,$$

so the deduction of the above inequality is correct. Therefore $f(2n) \leqslant 2f(n)+n$.

When we try to evaluate $f(2n+1)$, similar to the above discussions, we know that we only need to prove:

$$f(2n+1) \leqslant f(n)+f(n+1)+n.$$

Similarly, we discuss the cases $1 \leqslant j \leqslant n$, and separate the cases when j is even or odd.

When $j=2k$, $1 \leqslant k \leqslant \left[\frac{n}{2}\right]$, by induction hypothesis, we have the following:

$$f(j)+f(2n+1-j)+j$$
$$= f(2k)+f(2n+1-2k)+2k$$

$$= (2f(k) + k) + (f(n-k) + f(n-k+1) + n - k) + 2k$$
$$= (f(k) + f(n-k) + k) + (f(k) + f(n+1-k) + k) + n$$
$$\leqslant f(n) + f(n+1) + n.$$

When $j = 2k - 1$, $1 \leqslant k \leqslant \left[\frac{n+1}{2}\right]$, we have

$$f(j) + f(2n+1-j) + j$$
$$= f(2k-1) + f(2n-2k+2) + 2k - 1$$
$$= (f(k-1) + f(k) + k - 1) + (2f(n-k-1) + n - k + 1) + 2k - 1$$
$$= (f(k-1) + f(n-(k-1)) + k - 1) + (f(k) + f(n+1-k) + k) + n$$
$$\leqslant f(n) + f(n+1) + n.$$

Therefore, $f(2n+1) \leqslant f(n) + f(n+1) + n$.

Using all of the above, for any $n \in \mathbf{N}^*$, ① is always true.

Now we make use of ① to recursively calculate in turn and we can get the following:

$f(2) = 1$, $f(3) = 2$, $f(4) = 4$, $f(7) = 9$, $f(8) = 12$, $f(15) = 28$, $f(16) = 32$, $f(31) = 75$, $f(32) = 80$, $f(62) = 181$, $f(63) = 186$, $f(125) = 429$, $f(126) = 435$, $f(250) = 983$, $f(251) = 989$, $f(501) = 2222$, $f(1002) = 4945$, $f(2004) = 10\,892$.

The value in question is thus $f(2004) = 10\,892$.

Explanation. When we make guesses, it might be not very rigorous. But when we are doing the deductions or proofs, we must be very careful, otherwise, it is very easy to come to wrong conclusions and it is harmful as a scientific attitude and habit.

Example 2. For positive integers $k \geqslant 1$, we let $p(k)$ be the smallest prime number that cannot divide k. If $p(k) > 2$, we denote $q(k)$ to be the product of all prime numbers that are less than $p(k)$. If $p(k) = 2$, then we let $q(k) = 1$.

We define the sequence $\{x_n\}$ as follows: $x_0 = 1$, and

$$x_{n+1} = \frac{x_n p(x_n)}{q(x_n)},\ n = 0, 1, 2, \cdots.$$

Find all $n \in \mathbf{N}^*$ such that $x_n = 111\,111$.

Solution. We try calculating some initial values of x_n and is tabulated below.

If we write n as a binary number, then according to the above data, we may know that the number of 1s in the binary system for n is the number of prime numbers whose product is x_n. Taking a step further, we arrange the prime numbers from the smallest to biggest, assuming $p_0 < p_1 < p_2 < \cdots$. Checking it with the data in the above table, it is not hard to come to the conjecture below:

For any $n \in \mathbf{N}^*$, we suppose under the binary system,

$$n = 2^{r_1} + 2^{r_2} + \cdots + 2^{r_k},\ r_1 > r_2 > \cdots > r_k \geqslant 0.$$

That is, the binary number that corresponds to n has altogether $(r_1 + 1)$ digits, among which the elements on the digits $r_k + 1$, $r_{k-1} + 1$, $\cdots$, $r_1 + 1$ are 1s, on other digits the elements are all zeros. Then $x_n = p_{r_1} p_{r_2} \cdots p_{r_k}$, where p_{r_i} represents the $r_i + 1$th largest number among all prime numbers. ①

We are to prove the above conclusion by inducting with respect to n.

When $n = 1$, from $x_1 = 2 = p_0$ we know that ① holds.

Now we suppose the proposition is true for some n, i. e., $x_n = p_{r_1} p_{r_2} \cdots p_{r_k}$. Consider the case $n + 1$.

If $r_k \geqslant 1$, that is, the last digit of the binary expression of n is 0, then $n + 1 = 2^{r_1} + 2^{r_2} + \cdots + 2^{r_k} + 2^0$. At this time, x_n is an odd number, so $p(x_n) = 2$ and further, $q(x_n) = 1$. By the induction hypothesis, we know

$$x_{n+1} = \frac{x_n p(x_n)}{q(x_n)} = \frac{x_n \cdot p_0}{1} = p_{r_1} \cdots p_{r_k} p_0.$$

If $r_k = 0$, suppose i is the largest positive integer that makes $r_{i-1} \geqslant r_i + 2$, i.e., if we count from the second to last digit of the number in binary corresponding to n towards the left, among all the binary digits, only the $(r_i + 1)$th digit is the first, whose binary digit at the left-hand side contains at least one 0. That is,

$$n = \underset{\substack{r_1+1\text{th}\\ \text{digit}}}{1}\ 0\cdots0\ \underset{\substack{r_2+1\text{th}\\ \text{digit}}}{1}\ 0\cdots0\ \underset{\substack{r_{i-1}+1\text{th}\\ \text{digit}}}{1}\ 0\cdots0\ \underset{\substack{r_i+1\text{th}\\ \text{digit}}}{1}\ \underbrace{11\cdots11}_{r_i\text{ items}}.$$

At this time, $r_{k-j} = j$, where $0 \leqslant j \leqslant k - i$. Then

$$n + 1 = 2^{r_1} + 2^{r_2} + \cdots + 2^{r_{i-1}} + 2^{r_i+1}$$

(If i does not exist, then $n + 1 = 2^{r_1+1}$).

At this time, we know by induction hypothesis that $p(x_n) = p_{r_i+1}$, so

$$q(x_n) = p_0 p_1 \cdots p_{r_i} = p_0 p_1 \cdots p_{k-i} = p_{r_k} p_{r_{k-1}} \cdots p_{r_i}.$$

So

$$x_{n+1} = \frac{x_n \cdot p(x_n)}{q(x_n)} = \frac{p_{r_1} \cdots p_{r_{i-1}} p_{r_i+1} p_{r_i} \cdots p_{r_k}}{p_{r_i} \cdots p_{r_k}} = p_{r_1} \cdots p_{r_{i-1}} p_{r_i+1}.$$

So, ① is correct for $n + 1$, i.e., for any $n \in \mathbf{N}^*$, ① holds.

Now from

$$111\,111 = 3 \times 7 \times 11 \times 13 \times 37 = p_1 p_3 p_4 p_5 p_{11},$$

we can get the binary expression for the positive integer n which satisfies $x_n = 111\,111$ as follows,

$$n = 2^{11} + 2^5 + 2^4 + 2^3 + 2 = 2106.$$

So, the number n we look for is $n = 2106$.

Example 3. The integer sequence $\{a_n\}$ is defined as follows:

$$a_1 = 2,\ a_2 = 7,\ -\frac{1}{2} < a_{n+1} - \frac{a_n^2}{a_{n-1}} \leqslant \frac{1}{2},\ n = 2, 3, \cdots.$$

Find the explicit formula of the sequence $\{a_n\}$.

Solution. It is not easy to find a_n at first glance of the recursive formula given by the question. Can we get the linear recursive formula with constant coefficients from the condition? We may boldly guess $a_{n+1} = pa_n + qa_{n-1}$, where p, q are constants to be determined.

Let's try calculating the initial few terms and get $a_1 = 2$, $a_2 = 7$, $a_3 = 25$, $a_4 = 89$, $\cdots$. We may calculate the coefficients by using these initial terms and come to a conjecture: $a_{n+1} = 3a_n + 2a_{n-1}$, $n \geqslant 2$.

Next, we are to prove the above conjecture using mathematical induction.

The above conjecture is correct for $n = 2$, 3.

Suppose for $k \leqslant n$, we have $a_{k+1} = 3a_k + 2a_{k-1}$. Then for the case $k = n+1$, we have

$$\frac{a_{n+1}^2}{a_n} = \frac{a_{n+1}(3a_n + 2a_{n-1})}{a_n} = 3a_{n+1} + 2a_n + 2\left(\frac{a_{n+1}a_{n-1} - a_n^2}{a_n}\right).$$

Notice that

$$\begin{aligned}\left|2\left(\frac{a_{n+1}a_{n-1} - a_n^2}{a_n}\right)\right| &= \left|\frac{2a_{n-1}}{a_n}\right|\left|a_{n+1} - \frac{a_n^2}{a_{n-1}}\right| \\ &\leqslant \frac{1}{2}\left|\frac{2a_{n-1}}{a_n}\right|.\end{aligned}$$

By the induction hypothesis, we know $a_n > 2a_{n-1}$. So

$$\left|3a_{n+1} + 2a_n - \frac{a_{n+1}^2}{a_n}\right| < \frac{1}{2}.$$

Since a_{n+2} is an integer, and $\left|a_{n+2} - \dfrac{a_{n+1}^2}{a_n}\right| \leqslant \dfrac{1}{2}$, we get

$$\begin{aligned}&|a_{n+2} - (3a_{n+1} + 2a_n)| \\ = &\left|a_{n+2} - \frac{a_{n+1}^2}{a_n}\right| + \left|\frac{a_{n+1}^2}{a_n} - (3a_{n+1} + 2a_n)\right| \\ < &\frac{1}{2} + \frac{1}{2} = 1.\end{aligned}$$

So $a_{n+2} = 3a_{n+1} + 2a_n$. Then the conjecture is true for the case $k = n+1$.

Generalizing all of above, we know the sequence $\{a_n\}$ satisfies $a_1 = 2$, $a_2 = 7$, $a_n = 3a_{n-1} + 2a_{n-2}$, $n = 3, 4, \cdots$. By utilizing the characteristics equation of recursive sequences, we solve the linear recursive formula with constant coefficients and get

$$a_n = \frac{17 + 5\sqrt{17}}{68}\left(\frac{3 + \sqrt{17}}{2}\right)^n + \frac{17 - 5\sqrt{17}}{68}\left(\frac{3 - \sqrt{17}}{2}\right)^n.$$

Example 4. The function f: $\mathbf{N}^* \to \mathbf{N}^*$ is defined as follows: $f(1) = 1$, for $n \in \mathbf{N}^*$, the number $f(n+1)$ is the greatest positive integer m that satisfies the following conditions: There exists an arithmetic sequence $a_1, a_2, \cdots, a_m$ comprised of positive integers (here a sequence with less than 3 terms is also regarded as an arithmetic sequence), such that $a_1 < a_2 < \cdots < a_m = n$, and $f(a_1) = f(a_2) = \cdots = f(a_m)$. Prove that for any positive integer n, we have $f(4n+8) = n+2$.

Proof. The question does not require us to evaluate each function value at n. But if we look at the definition of f, only when each previous value of $f(n)$ is found can we easily find the next one.

We do calculations for initial values by using the definition of f so that we may know that:

$$f(1)=1,\ f(2)=1,\ f(3)=2,\ f(4)=1,\ f(5)=2,\ f(6)=2,$$
$$f(7)=2,\ f(8)=3,\ f(9)=1,\ f(10)=2,\ f(11)=2,\ f(12)=3,$$
$$f(13)=2,\ f(14)=3,\ f(15)=2,\ f(16)=4,\ f(17)=1,\ f(18)=3,$$
$$f(19)=2,\ f(20)=5,\ f(21)=1,\ f(22)=2,\ f(23)=2,\ f(24)=6,$$
$$f(25)=1,\ f(26)=4,\ f(27)=2,\ f(28)=7,\ f(29)=1,\ f(30)=4,$$
$$f(31)=2,\ f(32)=8,\ f(33)=1,\ f(34)=5,\ f(35)=2,\ f(36)=9,\ \cdots.$$

These listed values show that when $1 \leqslant n \leqslant 7$, we have $f(4n+8) = n+2$. Further, it pushes us to guess that when $n \geqslant 8$, we have

$$f(4n+1)=1;\ f(4n+2)=n-3;$$
$$f(4n+3)=2;\ f(4n+4)=n+1. \qquad ①$$

Next, we want to prove that when $n \geqslant 8$, ① is always true by inducting with respect to n.

When $n = 8$, we know that ① is true by using the listed values above.

Now we suppose that ① holds true for $8, 9, \cdots, n-1$. We consider the case $n (\geqslant 9)$.

By making use of the calculated values from $f(1)$ to $f(36)$ together with our induction hypothesis, we may know that $f(4n) = n$ is the maximum value among $f(1), f(2), \cdots, f(4n)$. So $f(4n+1) = 1$.

Now we study the items that are equal to 1 from $f(1)$ to $f(4n+1)$, *and we find* $f(17)=f(21)=\cdots=f(4n+1)=1$. Combined with the definition of f we come to $f(4n+2)\geqslant n-3$. On the other hand, for the arithmetic sequence $a_1<a_2<\cdots<a_m(=4n+1)$ with $4n+1$ as the ending term, if $f(a_1)=\cdots=f(a_m)=1$, then the common difference of this sequence $d\geqslant 4$, for if $d\leqslant 3$, then at least two of $f(4n-2)$, $f(4n-1)$, $f(4n)$, $f(4n+1)$ should be equal to 1; however, by the induction hypothesis, there is only $f(4n+1)=1$. If $d>4$, then by the induction hypothesis and the values shown from $f(1)$ to $f(36)$ we know that $d\geqslant 8$. So, $m\leqslant 1+\frac{(4n+1)-1}{8}<n-3$. Hence, $f(4n+2)=n-3$.

Then we evaluate the values from $f(1)$ to $f(4n+2)$, there is only $f(4n-12)=f(4n+2)=n-3$ (Here we made use of $n\geqslant 9$), therefore $f(4n+3)=2$.

Finally, by a similar manner of discussing the value of $f(4n+2)$, we may know that

$$f(4n+4)=n+1.$$

So, for any $n\in\mathbf{N}^*$ $(n\geqslant 8)$, ① always holds. Furthermore, for any $n\in\mathbf{N}^*$, we have $f(4n+8)=n+2$.

Explanation. To guess the result from a regular pattern, the number of initial values we need to calculate can be different for different problems. In this case, carefulness and confidence are both important.

Example 5. For any $n\in\mathbf{N}^*$, denote $\rho(n)$ to be a non-negative integer k that satisfies $2^k\mid n$ and $2^{k+1}\nmid n$. The sequence $\{x_n\}$ is defined as follows:

$$x_0=0,\ \frac{1}{x_n}=1+2\rho(n)-x_{n-1},\ n=1,\ 2,\ \cdots.$$

Prove that every non-negative rational number will appear exactly once in the sequence x_0, x_1, $\cdots$.

Proof. If we write $x_n = \frac{p_n}{q_n}$ $(p_n, q_n \in \mathbf{N}^*, (p_n, q_n) = 1)$, then the condition will be

$$\frac{q_n}{p_n} = (1 + 2\rho(n)) - \frac{p_{n-1}}{q_{n-1}}.$$

It is the most convenient to eliminate the denominator when $p_n = q_{n-1}$. This guess brings out the following proof.

Define the sequence $\{y_n\}$ as follows:

$$y_1 = y_2 = 1,\ y_{n+2} = (1 + 2\rho(n))y_{n+1} - y_n,\ n = 1, 2, \cdots.$$

We come to the following conclusions in turn.

Conclusion 1. For any $n \in \mathbf{N}^*$, we always have $x_n = \frac{y_n}{y_{n+1}}$.

We will prove it by inducting with respect to n. The inductive step for induction could proceed as follows.

$$\begin{aligned}\frac{1}{x_{n+1}} &= 1 + 2\rho(n+1) - x_n = 1 + 2\rho(n+1) - \frac{y_n}{y_{n+1}} \\ &= \frac{1}{y_{n+1}}((1 + 2\rho(n+1))y_{n+1} - y_n) \\ &= \frac{y_{n+2}}{y_{n+1}}.\end{aligned}$$

So $x_{n+1} = \frac{y_{n+1}}{y_{n+2}}$.

Conclusion 2. For any $n \in \mathbf{N}^*$, we always have $y_{2n+1} = y_{n+1} + y_n$, $y_{2n} = y_n$.

We will induct with respect to n. As a matter of fact, if Conclusion 2 is true for n, then

$$\begin{aligned}y_{2n+2} &= (1 + 2\rho(2n+1))y_{2n+1} - y_{2n} = y_{2n+1} - y_n = y_{n+1}; \\ y_{2n+3} &= (1 + 2\rho(2n+2))y_{2n+2} - y_{2n+1} \\ &= (1 + 2(1 + \rho(n+1)))y_{2n+2} - y_{2n+1} \\ &= 2y_{n+1} + (1 + \rho(n+1))y_{n+1} - (y_{n+1} + y_n) \\ &= y_{n+1} + (1 + \rho(n+1))y_{n+1} - y_n \\ &= y_{n+1} + y_{n+2}.\end{aligned}$$

By this and the initial conditions, we may know that Conclusion 2 is correct.

From Conclusion 2 combined with mathematical induction method, it is easy to prove for any $n \in \mathbf{N}^*$, we always have $(y_n, y_{n+1}) = 1$.

Conclusion 3. For any $p, q \in \mathbf{N}^*$, $(p, q) = 1$, there exists a unique $n \in \mathbf{N}^*$ such that $(p, q) = (y_n, y_{n+1})$.

We will prove by inducting with respect to $p+q$. When $p+q=2$, $p=q=1$. At this time $(p, q) = (y_1, y_2)$, while from Conclusion 2 we know that when $n \geqslant 2$, at least one from y_n and y_{n+1} is bigger than 1. So, $(y_n, y_{n+1}) \neq (y_1, y_2)$. Therefore, Conclusion 3 is true for $p+q=2$.

Now we suppose Conclusion 3 is true for all positive integer pairs (p, q) satisfying $p+q<m$ $(m \geqslant 3, m \in \mathbf{N}^*)$ and $(p, q) = 1$. Consider the case $p+q=m$. At this time $p \neq q$, and we may discuss under two cases, $p<q$ and $p>q$.

Case 1. $p<q$. From $(p, q)=1$ we know that $(p, q-p)=1$, however, $(q-p)+p=q<m$. By induction hypothesis, there exists a unique $n \in \mathbf{N}^*$, such that $(p, q-p) = (y_n, y_{n+1})$. Then $(p, q) = (y_n, y_n+y_{n+1}) = (y_{2n}, y_{2n+1})$. (Here we used Conclusion 2.)

On the other hand, if there exists $k<l$, $k, l \in \mathbf{N}^*$, such that $(p, q) = (y_k, y_{k+1}) = (y_l, y_{l+1})$, then $y_k = y_l$, $y_{k+1} = y_{l+1}$. At this time, if k and l are both even numbers, then by Conclusion 2 we know that $(p, q-p)$ has two different expressions, which is contradictory to the induction hypothesis. But when k is odd, $y_k > y_{k+1}$, and this is contradictory with $p<q$. So k is an even number. By the same reasoning l is an even number. Therefore, there exists only one $n \in \mathbf{N}^*$ such that $(p, q) = (y_n, y_{n+1})$.

Case 2. $p>q$. Discuss in a similar manner as Case 1.

By all above, Conclusion 3 holds.

By Conclusion 1 and 3, and $x_0=0$, we may know that the proposition is true.

15 Problems Regarding Existence with Sequences

Problems regarding existence appear on all branches of mathematics. They also showed up in previous sections of this book. Here we dedicate one section to discuss the existence problems of sequences with the aim of stressing it and bringing it attention. We discuss the ways to handle this kind of questions in the form of examples.

Example 1. Suppose a, b are integers that are bigger than 2. Prove that there exist a positive integer k and a finite sequence $n_1, n_2, \cdots, n_k$ of positive integers such that $n_1 = a$, $n_k = b$, while for $1 \leqslant i \leqslant k-1$, it is always true that $(n_i + n_{i+1}) \mid n_i n_{i+1}$.

Proof. We use the notation "$a \sim b$" to imply that the positive integers a, b can be "connected" by the above-like sequence, then "if $a \sim b$ holds, then $b \sim a$ also holds".

A natural idea is to prove: for any two adjacent positive integers (both bigger than 2), they are "connectable". We can meet this objective by use of the following two conclusions.

Conclusion 1. For any $n \in \mathbf{N}^*$, $n \geqslant 3$, it is always true that $n \sim 2n$.

The following sequence shows that Conclusion 1 holds.

$$n,\ n(n-1),\ n(n-1)(n-2),\ n(n-2),\ 2n.$$

Conclusion 2. For any $n \in \mathbf{N}^*$, $n \geqslant 4$, it is always true that $n \sim n-1$.

We use the sequence

$$n,\ n(n-1),\ n(n-1)(n-2),\ n(n-1)(n-2)(n-3),\ 2(n-1)(n-2).$$

Combining Conclusion 1 from which we know that $2(n-1)(n-2) \sim (n-1)(n-2)$, with $(n-1)(n-2)+(n-1) = (n-1)^2$ is a divisor of $(n-1)(n-2) \cdot (n-1)$, we know that Conclusion 2 holds.

For integers a, b that are greater than 2, without loss of generality, suppose $a \leqslant b$. If $a = b$, then using $a \sim a+1 \sim b(=a)$ we know that the proposition holds. If $a < b$, then using $a \sim a+1 \sim a+2 \sim \cdots \sim b$ we also know that the proposition holds.

Explanation. The key to solving this problem was the direct construction of Conclusion 1 and Conclusion 2. This is the most natural way of thinking when dealing with problems of existence.

Example 2. Suppose $m \in \mathbf{N}^*$. Tell whether there exists a polynomial $f(x)$ of degree n with integer coefficients such that for any $n \in \mathbf{Z}$, any two terms from the sequence $\{a_k\}$ defined by the following method are coprime: $a_1 = f(n)$, $a_{k+1} = f(a_k)$, $k = 1, 2, \cdots$.

Solution. When $m = 1$, this kind of polynomial does not exist.

As a matter of fact, if there exists a function $f(x) = ax + b$ that meets the requirements, without loss of generality, suppose $a > 0$. Then for any $n \in \mathbf{Z}$, we have

$$a_k = a^k \cdot n + (a^{k-1} + \cdots + 1)b. \qquad ①$$

This conclusion could be reached by inducting with respect to k.

If $b = 0$, then for any positive integer n that is bigger than 1, we may know from ① that every term of the sequence $\{a_k\}$ is a multiple of n; therefore, there are no terms that are coprime.

If $b \neq 0$, since a is a positive integer, we know that there exists a $k \in \mathbf{N}^*$ such that $|(a^{k-1} + \cdots + 1)b| > 1$. Denote $c = (a^{k-1} + \cdots + 1)b$. We take n as a prime factor of $|c|$, then the a_k corresponding to this n is a multiple of n. From ① we know that

$$\begin{aligned} a_{2k} &= a^{2k} \cdot n + (a^{2k-1} + \cdots + 1)b \\ &= a^{2k} \cdot n + (a^k + 1) \cdot (a^{k-1} + \cdots + 1)b \\ &= a^{2k} \cdot n + (a^k + 1)c. \end{aligned}$$

So n is also a divisor of a_{2k}, which makes a_k and a_{2k} not coprime.

So, when $m = 1$, there does not exist a polynomial with integer coefficients that meets the requirements.

Next, we are to prove that when $m \geqslant 2$, this kind of polynomial always exists.

We will prove: When $f(x) = x^{m-1}(x-1)+1$, for any $n \in \mathbf{Z}$, any two terms from the corresponding sequence $\{a_k\}$ are coprime.

We notice that for any $k \in \mathbf{N}^*$, we have $a_{k+1} = a_k^{m-1}(a_k - 1) + 1 \equiv 1 \pmod{a_k}$, and also,

$$a_{k+2} = a_{k+1}^{m-1}(a_{k+1} - 1) + 1 \equiv 1^{m-1} \cdot 0 + 1 = 1 \pmod{a_k}.$$

Following this, we may know by mathematical induction that, for any positive integer $t > k$, it is always true that $a_t \equiv 1 \pmod{a_k}$. So, any two terms from the sequence $\{a_k\}$ are coprime.

From all above, when $m = 1$, this kind of polynomial does not exist. When $m \geqslant 2$, we can always find such polynomials.

Explanation. For the case $m \geqslant 2$, we take an arbitrary polynomial $g(x)$ of degree $m-2$ with integer coefficients.

Let $f(x) = x(x-1)g(x)+1$. By a similar method as above, we may prove that for $n \in \mathbf{Z}$, any two terms from the corresponding sequence $\{a_k\}$ are coprime.

Example 3. Suppose q is a given real number which satisfies $\frac{1+\sqrt{5}}{2} < q < 2$. The sequence $\{p_n\}$ is defined as follows: If the binary expression of a positive integer n is $n = 2^m + a_{m-1} \cdot 2^{m-1} + \cdots + a_1 \cdot 2 + a_0$, where $a_i \in \{0, 1\}$, then

$$p_n = q^m + a_{m-1} \cdot q^{m-1} + \cdots + a_1 \cdot q + a_0.$$

Prove that there exist infinitely many positive integers k such that there does not exist a positive integer l which satisfies $p_{2k} < p_l < p_{2k+1}$.

Proof. For $m \in \mathbf{N}^*$, suppose under the binary system,

$$2k = (\underbrace{10\cdots10}_{m\text{ items}})_2.$$

We prove that there doesn't exist an $l \in \mathbf{N}^*$, such that $p_{2k} < p_l < p_{2k+1}$.

As a matter of fact, for this kind of $k \in \mathbf{N}^*$, we have

$$p_{2k} = q^{2m-1} + q^{2m-3} + \cdots + q, \quad p_{2k+1} = p_{2k} + 1.$$

If there exists an $l \in \mathbf{N}^*$, such that $p_{2k} < p_l < p_{2k+1}$, then we suppose the binary expression for l is $l = \sum_{i=0}^{t} a_i \cdot 2^i$, where $a_i \in \{0, 1\}$, $a_t = 1$, then $p_l = \sum_{i=0}^{t} a_i \cdot q^i$.

(1) If $m = 1$, then $q < p_l < q + 1$. At this time, if $t \geqslant 2$, then $p_l \geqslant q^2 > q + 1$ $\left(\text{since } \frac{1+\sqrt{5}}{2} < q < 2, \text{ we have } q + 1 < q^2\right)$. This causes a contradiction. If $t = 1$, then $p_l = q$, or $q + 1$, also contradictory.

(2) Suppose when it is $m - 1$ $(m \geqslant 2)$, we may come to a contradiction. Consider the case m.

If $t \geqslant 2m$, then

$$\begin{aligned} p_l &\geqslant q^{2m} \geqslant q^{2m-1} + q^{2m-2} \geqslant q^{2m-1} + q^{2m-3} + q^{2m-4} \geqslant \cdots \\ &\geqslant q^{2m-1} + \cdots + q + 1 = p_{2k+1}. \end{aligned}$$

Contradiction.

If $t \leqslant 2m - 2$, then

$$\begin{aligned} p_l &\leqslant q^{2m-2} + q^{2m-3} + \cdots + 1 \\ &= (q^{2m-2} + q^{2m-3}) + (q^{2m-4} + q^{2m-5}) + \cdots + (q^2 + q) + 1 \\ &\leqslant q^{2m-1} + q^{2m-3} + \cdots + q^3 + 1 \\ &< q^{2m-1} + \cdots + q^3 + q = p_{2k}. \end{aligned}$$

Contradiction.

During the above reasoning process, we have used $q^{i+2} \geqslant q^{i+1} + q^i$, $i = 0, 1, 2, \cdots$.

So $t = 2m - 1$. At this time, we denote $l' = l - 2^{2m-1} = \sum_{i=0}^{t-1} a_i \cdot 2^i$. Furthermore, we have $p_{l'} = p_l - q^{2m-1}$. Then, from $p_{2k} < p_l < p_{2k+1}$, we know that

$$p_{2(k-1)} = q^{2m-3} + \cdots + q^3 + q < p_{l'} < p_{2(k-1)} + 1.$$

This is inconsistent with the induction hypothesis.

From all of above, the proposition holds true.

Example 4. State whether there exists a sequence $\{a_n\}$ of positive integers such that every positive integer appears only once in this sequence, and that for any $k \in \mathbf{N}^*$, we have $k \mid (a_1 + \cdots + a_k)$.

Solution. There exists such a sequence.

We construct such a sequence by a recursive method. Take $a_1 = 1$, and we suppose $a_1, a_2, \cdots, a_m$ (all different) are already chosen. Let t be the least positive integer that doesn't show up in $a_1, \cdots, a_m$. Since $(m+1, m+2) = 1$, then by using Chinese Remainder Theorem, we may know that there exist infinitely many positive integers r such that (Denote $s = a_1 + \cdots + a_m$)

$$\begin{cases} s + r \equiv 0 \pmod{m+1}, \\ s + r + t \equiv 0 \pmod{m+2}. \end{cases}$$

Take such an r which makes $r > \max\{a_1, \cdots, a_m, t\}$. Let $a_{m+1} = r$, $a_{m+2} = t$. A sequence defined like above will meet all requirements.

Explanation. The recursive method applied to solve problems of existence, in essence, is a technique of direct construction. The sequence we defined could be $1, 3, 2, 10, 4, \cdots$. Every time we have two more terms and this practice will make sure that the sequence covers all positive integers without repetition.

Example 5. A sequence $\{a_n\}$ of all integers satisfies the following conditions. For any subscript $k \geqslant 2$, we have $0 \leqslant a_k \leqslant k-1$, and also $a_1 + \cdots + a_k \equiv 0 \pmod{k}$. Prove that no matter what initial value a_1 is chosen, there exists a positive integer m, such that for this sequence, starting from the mth term, all the terms are constants.

Proof. The starting point is to prove that for any $a_1 \in \mathbf{Z}$, there exists a subscript k such that $a_1 + \cdots + a_k = dk$, where $0 \leqslant d < k$. ①

If the above conclusion is successfully proved, then $a_1 + \cdots + a_k + d = d \cdot (k+1)$. Note that a_{k+1} is the only integer that satisfies $a_1 + \cdots + a_{k+1} \equiv 0 \pmod{k+1}$ in the set $\{0, 1, 2, \cdots, k\}$. Then $a_{k+1} = d$. According to this and reason recursively, we can prove that when $n \geqslant$

$k+1$, we always have $a_n = d$.

Now we will prove that ① holds true. If this is not true, suppose there exists an a_1 such that the subscript k that satisfies ① does not exist. Since when $a_1 < 0$, if for the sequence $\{a_n\}$, there is no term starting from which all terms are all zeros, then there will be infinitely many terms in the sequence $\{a_n\}$ that are positive integers. Hence, there exists an $m \in \mathbf{N}^*$ such that $a_1 + \cdots + a_m \geqslant 0$. Hence, without loss of generality, we may suppose $a_1 > 0$. (Attention: If $a_1 = 0$, then we can know that for any $n \in \mathbf{N}^*$, we always have $a_n = 0$.) At this time, for any $m \in \mathbf{N}^*$, it is always true that $a_1 + a_2 + \cdots + a_m > 0$.

By the condition $a_1 + \cdots + a_m \equiv 0 \pmod{m}$, we may suppose $a_1 + \cdots + a_m = d_m \cdot m$. Combining the counter-hypothesis that there is no subscript k that satisfies ①, we know that for any $m \in \mathbf{N}^*$, we have $d_m \geqslant m$, hence $a_1 + \cdots + a_m \geqslant m^2$. When $m \geqslant 2$, we have $a_m \leqslant m-1$, then

$$m^2 \leqslant a_1 + \cdots + a_m \leqslant a_1 + 1 + 2 + \cdots + (m-1) = a_1 + \frac{m(m-1)}{2}.$$

This leads to $a_1 \geqslant \frac{m(m+1)}{2}$, and it is not always true for all $m \in \mathbf{N}^*$. The contradiction we arrived at shows that ① is true.

From all above, we know that the proposition holds true.

Explanation. Proof by Contradiction (and Drawer Principle or the Pigeonhole Principle) is a basic indirect method to solve problems of existence. It is more common to use this kind of method to deal with non-existence problems.

Example 6. The sequence $\{a_n\}$ is defined as follows: If a positive integer n under the binary system, the number 1 appears for even number of times, then let $a_n = 0$, otherwise let $a_n = 1$. Prove that: there do not exist positive integers k, m such that for any $j \in \{0, 1, 2, \cdots, m-1\}$, it is always true that

$$a_{k+j} = a_{k+m+j} = a_{k+2m+j}. \qquad ①$$

Proof. By making use of the definition of $\{a_n\}$ we know that

$$\begin{cases} a_{2n} \equiv a_n \pmod 2, \\ a_{2n+1} \equiv a_{2n} + 1 \equiv a_n + 1 \pmod 2. \end{cases} \quad ②$$

If there exist k, $m \in \mathbf{N}^*$ such that for $j \in \{0, 1, \cdots, m-1\}$, ① is always true, then by Principle of the Minimum Natural Number (the Well-ordering Principle), we may suppose (k, m) is the positive integer pair that makes the sum $k+m$ the smallest.

Case 1. m is an even number. Then we suppose $m = 2t$, $t \in \mathbf{N}^*$.

If k is an even number, then we take $j = 0, 2, \cdots, 2(t-1)$ in ①, where $0 \leqslant \frac{j}{2} \leqslant t-1$, and

$$a_{k+j} = a_{k+m+j} = a_{k+2m+j}.$$

From ② we have $a_{\frac{k}{2}+\frac{j}{2}} = a_{\frac{k}{2}+t+\frac{j}{2}} = a_{\frac{k}{2}+2t+\frac{j}{2}}$. This shows that $\left(\frac{k}{2}, \frac{m}{2}\right)$ is also a positive integer pair that makes ① true for $0 \leqslant j \leqslant \frac{m}{2} - 1$, contradicting the previous conclusion that $k+m$ is the smallest.

If k is an odd number, then we take $j = 1, 3, \cdots, 2t-1$ in ①. By similar discussions as above, we have

$$a_{\frac{k+1}{2}+\frac{j-1}{2}} = a_{\frac{k+1}{2}+t+\frac{j-1}{2}} = a_{\frac{k+1}{2}+2t+\frac{j-1}{2}},$$

which shows that $\left(\frac{k+1}{2}, \frac{m}{2}\right)$ also makes ① true for $0 \leqslant j \leqslant \frac{m}{2} - 1$, contradicting the previous conclusion that $k+m$ is the smallest.

Case 2. m is an odd number.

When $m = 1$, it is required that $a_k = a_{k+1} = a_{k+2}$. At this time if k is an even number, then $a_{2n} = a_{2n+1} \equiv a_{2n} + 1 \pmod 2$, contradicting what we know. If k is an odd number, then we let $k = 2n+1$, and we have $a_{2n+2} = a_{2n+3} \equiv a_{2n+2} + 1 \pmod 2$, also contradicting what we know.

When $m \geqslant 3$, let $j = 0, 1, 2$ in ①, then we have

$$\begin{cases} a_k = a_{k+m} = a_{k+2m}, & ③ \\ a_{k+1} = a_{k+m+1} = a_{k+2m+1}, & ④ \\ a_{k+2} = a_{k+m+2} = a_{k+2m+2}. & ⑤ \end{cases}$$

If k is an even number, let $k = 2n$, $m = 2t + 1$, then from ② we know that $a_{k+1} \neq a_k$, $a_{k+m+1} \neq a_{k+m+2}$. Then by combining ③, ④ and ⑤ we can know that

$$a_k = a_{k+m+2} = a_{k+2}. \quad ⑥$$

(Attention: every term in the sequence that we use is either 0 or 1.)

Now, if n is an even number, let $n = 2t$, then

$$a_{k+2} = a_{4t+2} = a_{2t+1} \equiv a_{2t} + 1 \equiv a_{4t} + 1 \equiv a_k + 1 \pmod 2,$$

contradicting ⑥; If n is an odd number, then from the fact that m is an odd number we get $k + 2m \equiv 0 \pmod 4$. By similar discussions we have $a_{k+2m} \neq a_{k+2m+2}$, combining ③, ⑤ and ⑥ we come to a contradiction.

If k is an odd number, combining the fact that m is an odd number and from ② we know that $a_{k+m} \neq a_{k+m+1}$, $a_{k+1} \neq a_{k+2}$. By referring to ③, ④ and ⑤, we have

$$a_k = a_{k+m} = a_{k+2}. \quad ⑦$$

Now, if $k \equiv 1 \pmod 4$, then we can come to a contradiction from $a_k = a_{k+2}$ in ⑦. If $k \equiv 3 \pmod 4$, then by the fact that m is an odd number we can know that $k + 2m \equiv 1 \pmod 4$, so $a_{k+2m} \neq a_{k+2m+2}$, i.e., $a_k \neq a_{k+2}$, contradicting ⑦.

From all above, the proposition holds true.

Exercise Set 2

1. Suppose S is a set with 2011 elements, and N is an integer satisfying $0 \leqslant N \leqslant 2^{2011}$.

Prove that we can dye every subset of S into black or white such that

(1) the union of any two white subsets is still white.

(2) the union of any two black subsets is still black.

(3) there are exactly N subsets that are white.

2. Place 2048 numbers on a circle, which are all $+1$ or -1. Now we multiply each number by its right neighbor and we replace the original number by the product we get, giving a circle of new numbers. Prove that after a finite number of operations like this, all numbers on the circle will become $+1$.

3. Suppose $x_1, \cdots, x_n$ are any real numbers. Prove that:

$$\sum_{i=1}^{n} \frac{x_i}{1+x_1^2+\cdots+x_i^2} < \sqrt{n}.$$

4. Suppose $n \in \mathbf{N}^*$, and the complex numbers $z_1, \cdots, z_n; \omega_1, \cdots, \omega_n$ satisfy for any vector $(\varepsilon_1, \cdots, \varepsilon_n)$, $\varepsilon_i \in \{-1, 1\}$, $i=1, 2, \cdots, n$. it is always true that

$$|\varepsilon_1 z_1 + \cdots + \varepsilon_n z_n| \leqslant |\varepsilon_1 \omega_1 + \cdots + \varepsilon_n \omega_n|.$$

Prove that $|z_1|^2 + \cdots + |z_n|^2 \leqslant |\omega_1|^2 + \cdots + |\omega_n|^2$.

5. Suppose $P(x_1, x_2, \cdots, x_n)$ is a polynomial with n variables. We replace all the variables in P by $+1$ or -1. If there are even -1 s, then the value of P is positive. If there are odd -1 s, then P is negative. Prove that P is a polynomial with at least degree n. (Namely, there is a term in P such that the sum of all degrees of the variables in this term is no less than n.)

6. Suppose $a_1, \cdots, a_n$ is a sequence of all non-negative real numbers(not all of them are zeros). Define

$$m_k = \max_{1\leqslant i\leqslant k} \frac{a_{k-i+1}+a_{k-i+2}+\cdots+a_k}{i}, \quad k=1, 2, \cdots, n.$$

Prove that for any positive real number μ, the number of subscripts k that satisfy $m_k > \mu$ is less than $\dfrac{a_1+a_2+\cdots+a_n}{\mu}$.

7. (Jensen's Inequality) Suppose $f(x)$ is a convex function on $[a, b]$ $\left(\text{namely, for any } x, y \in [a, b], \text{ we always have } f\left(\frac{x+y}{2}\right) \leqslant \frac{1}{2}(f(x)+f(y))\right)$.

Prove that for any n numbers $x_1, \cdots, x_n \in [a, b]$, it is always true that

$$f\left(\frac{x_1+\cdots+x_n}{n}\right)\leqslant\frac{1}{n}(f(x_1)+\cdots+f(x_n)).$$

8. Suppose the real numbers x_1, $\cdots$, x_n satisfy $x_1+\cdots+x_n=1$, where $n\in\mathbf{N}^*$, $n\geqslant 2$. Prove that

$$\prod_{k=1}^{n}\left(1+\frac{1}{x_k}\right)\geqslant\prod_{k=1}^{n}\left(\frac{n-x_k}{1-x_k}\right).$$

9. The Fibonacci sequence $\{F_n\}$ satisfies: $F_1=F_2=1$, $F_{n+2}=F_{n+1}+F_n$. Prove that

$$\sum_{i=1}^{n}\frac{F_i}{2^i}<2.$$

10. Find the smallest positive integer k such that there exist at least two sequences $\{a_n\}$ of positive integers which satisfy the following conditions:

(1) For any positive integer n, we have $a_n\leqslant a_{n+1}$;

(2) For any positive integer n, we have $a_{n+2}=a_{n+1}+a_n$;

(3) $a_9=k$.

11. The Fibonacci Sequence $\{F_n\}$ is defined as follows: $F_1=F_2=1$, $F_{n+2}=F_{n+1}+F_n$, $n=1, 2, \cdots$., Find all positive integer pairs (k, m), $m>k$, such that the sequence $\{x_n\}$ defined below contains 1:

$$x_1=\frac{F_k}{F_m},\ x_{n+1}=\begin{cases}\dfrac{2x_n-1}{1-x_n}, & \text{if } x_n\neq 1,\\ 1, & \text{if } x_n=1.\end{cases}\quad(n=1, 2, \cdots)$$

12. Mr. Zhang takes randomly a number from $\{1, 2, \cdots, 144\}$. Mr. Wang wants to know the number Mr. Zhang got with the following game: Mr. Wang takes a subset M from $\{1, 2, \cdots, 144\}$ and then asks Mr. Zhang whether the number he took belongs to M. If the answer is a Yes, then Mr. Wang will pay Mr. Zhang 2 RMB. If the answer is a No, then Mr. Wang will pay Mr. Zhang 1 RMB. What is the minimum amount of money Mr. Wang needs to pay so that it is guaranteed he knows the number Mr. Zhang got?

13. The Fibonacci sequence $\{F_n\}$ satisfies $F_1=F_2=1$, $F_{n+2}=$

$F_{n+1}+F_n$, $n=1, 2, \cdots$.

Prove that for any positive integer m, there exists a subscript n, such that

$$m \mid (F_n^4 - F_n - 2).$$

14. We call an infinite sequence of positive integers an F-Sequence, given that from the 3rd term, every term in this sequence is equal to the sum of the two terms immediately before it. Is it possible to decompose the set of positive integers into the union of

(1) finite number of (2) infinite number of F-Sequences?

15. Suppose the integers k, a_1, $\cdots$, a_n satisfy $0<a_n<a_{n-1}<\cdots<a_1\leqslant k$, and that for any $1\leqslant i, j\leqslant n$, it is true $[a_i, a_j]\leqslant k$.

Prove that for any $i\in\{1, 2, \cdots, n\}$, it is true that $ia_i\leqslant k$.

16. Suppose that $a_0<a_1<\cdots<a_n$, a_0, $\cdots$, a_n are all positive integers. Prove that

$$\frac{1}{[a_0, a_1]}+\frac{1}{[a_1, a_2]}+\cdots+\frac{1}{[a_{n-1}, a_n]}\leqslant 1-\frac{1}{2^n}.$$

17. Define the sequence $\{u_n\}_{n=0}^{+\infty}$ as: $u_0=0$, $u_1=1$, and that for any $n\in\mathbf{N}^*$, the number u_{n+1} is the smallest positive integer that meets the following conditions:

(1) for any $n\in\mathbf{N}^*$, $u_{n+1}>u_n$;

(2) there are no three numbers from the sequence u_0, u_1, $\cdots$, u_{n+1} that form an arithmetic sequence.

Find the value of u_{100}.

18. The positive integers a, b, n $(b>1)$ satisfy $(b^n-1)\mid a$. Prove that under base b, within the expression for number a, there will be at least n non-zero numbers.

19. Suppose $n\in\mathbf{N}^*$, $n>1$, and denote $h(n)$ to be the biggest prime factor of n. Prove that there exist infinitely many $n\in\mathbf{N}^*$ such that $h(n)<h(n+1)<h(n+2)$.

20. Suppose $n\in\mathbf{N}^*$, $n>1$, and denote $w(n)$ to be the number of distinct prime factors of n. Prove that there exist infinitely many $n\in$

$\mathbf{N}^*$ such that $w(n) < w(n+1) < w(n+2)$.

21. We use a_n to express the sum of the initial n prime numbers.

Prove that for any $n \in \mathbf{N}^*$, there is at least one perfect square number within the interval $[a_n, a_{n+1}]$.

22. Prove that for any positive odd number, we may always find a positive integer such that the product of them, under the decimal system, has every digit being odd numbers.

23. Denote $A = \{x \mid x \in \mathbf{N}^*$, x has all its digits not being zero under the decimal system, and $s(x) \mid x\}$, where $s(x)$ denotes the sum of all digits of x.

(1) Prove that there exist infinitely many numbers in A, whose expressions under the decimal system have equal times of appearances for the digits 1, 2, ⋯, 9.

(2) Prove that for any $k \in \mathbf{N}^*$, there is one term in A which is exactly a positive integer with k digits.

24. Tell whether there exists an infinite sequence comprised of positive integers such that

(1) Every term is not a multiple of any other term.

(2) Any two terms from the sequence are not mutually prime, but there is no positive integer greater than 1 that can divide each term of the sequence.

25. Suppose p is an odd prime number, and $a_1, a_2, \cdots, a_{p-2}$ is a sequence of positive integers satisfying for any $k \in \{1, 2, \cdots, p-2\}$, it is always true that $p \nmid a_k(a_k^k - 1)$. Prove that we can take several numbers from $a_1, a_2, \cdots, a_{p-2}$ such that their product $\equiv 2 \pmod{p}$.

26. Suppose $f: \mathbf{N}^* \to \mathbf{N}^*$ is a one-on-one correspondence.

(1) Prove that there exist positive integers a, d, such that

$$f(a) < f(a+d) < f(a+2d);$$

(2) For any positive integer m that is no less than 5, tell whether there must exist positive integers a, d, such that $f(a) < f(a+d) < \cdots < f(a+md)$.

27. Prove that for any real numbers $\alpha \in (1, 2]$, there exists a unique sequence $\{n_k\}$ of positive integers such that $n_k^2 \leqslant n_{k+1}$, and

$$\alpha = \lim_{m \to +\infty} \prod_{k=1}^{m} \left(1 + \frac{1}{n_k}\right).$$

28. Suppose m is a given positive integer, and that every term of the sequence $\{a_n\}$ is a positive integer, and for any positive integer n, it is always true $0 < a_{n+1} - a_n \leqslant m$.

Prove that there exist infinitely many pairs of positive integers (p, q), such that $p < q$, and $a_p \mid a_q$.

29. Suppose S is a set of non-negative integers. We use $r_s(n)$ to denote the number of pairs of ordered pairs (s_1, s_2) that satisfy the following conditions: $s_1, s_2 \in S$, $s_1 \neq s_2$, and $s_1 + s_2 = n$.

Discuss whether we can partition the set of non-negative integers into two sets A and B, such that for any non-negative integer n, it is always true that $r_A(n) = r_B(n)$.

30. Prove that any integer bigger than 1 can be expressed as the form of a sum of a finite number of positive integers that satisfy the following conditions:

(1) the prime factors of each addend are either 2 or 3;

(3) neither of any two addends is a multiple of the other.

31. The functions f, g: $\mathbf{N}^* \to \mathbf{N}^*$ are defined where f is a surjective mapping, while g is an injective mapping. For any positive integer n, it is always true that $f(n) \geqslant g(n)$. Prove that for any positive integer n, it is always true $f(n) = g(n)$.

32. Does there exist a sequence $\{a_n\}$ of integers such that $0 = a_0 < a_1 < a_2 < \cdots$, and that it meets the following two conditions:

(1) Every positive integer can be expressed in the form of $a_i + a_j$ (i, $j \geqslant 0$ and they can be the same value.);

(2) For any positive integer n, it is true that $a_n > \frac{n^2}{16}$.

Solutions to Exercises

Solutions to Exercise Set 1

1. We induct with respect to n, the number of elements of this non-empty finite set. Denote this set to be S_n. When $n = 1$, its subsets can be arranged as $\varnothing$, S_1 and the requirement is met. Now suppose that the statement is true for n, i.e., the subsets of S_n can be permuted as A_1, A_2, $\cdots$, A_{2^n}, such that the number of elements for consecutive sets differs by 1. Consider $S_{n+1} = \{a_1, \cdots, a_{n+1}\}$. For its subset of n elements $S_n = \{a_1, \cdots, a_n\}$, we have a permutation A_1, $\cdots$, A_{2^n}, which are all subsets of S_n, which meets the requirement according to the induction hypothesis. Then, the following permutation:

$$A_1, \cdots, A_{2^n}, A_{2^n} \cup \{a_{n+1}\}, \cdots, A_1 \cup \{a_{n+1}\}.$$

is a permutation of all subsets of S_{n+1} that meets the requirement.

Explanation. Within the permutation of subsets we constructed here, the number of elements from any two consecutive subsets differs exactly by one. This is even stronger than what's required.

2. When $k = 0$, the statement is obviously true. For the case when $k > 0$, the conclusion to be proved is equivalent to $\frac{a_k}{k} \leqslant \frac{a_n}{n}$, and this is the corollary of the following statement: for $k \geqslant 0$, we always have

$$(k+1)a_k \leqslant ka_{k+1}. \qquad ①$$

We induct with respect to k to prove that ① holds true: When $k = 0$, from $a_0 = 0$ we know that ① holds true. Now we suppose ① is true for k, then we know from the conditions that

$$(k+2)a_{k+1} = 2(k+1)a_{k+1} - ka_{k+1} \leqslant 2(k+1)a_{k+1} - (k+1)a_k$$
$$= (k+1)(2a_{k+1} - a_k) \leqslant (k+1)a_{k+2}.$$

So, ① is also true for $k+1$. The proposition is proved.

3. When $n=1$, $a_1^2 \leqslant a_1 - a_2 < a_1$, so $a_1 < 1$, and at the same time, $a_2 \leqslant a_1 - a_1^2 = \frac{1}{4} - \left(a_1 - \frac{1}{2}\right)^2 \leqslant \frac{1}{4} < \frac{1}{2}$. So the proposition holds true for $n = 1, 2$. Now we suppose the proposition holds true for $n(\geqslant 2)$, then $a_{n+1} \leqslant a_n - a_n^2 = \frac{1}{4} - \left(\frac{1}{2} - a_n\right)^2$. Note that by induction hypothesis $a_n < \frac{1}{n} \leqslant \frac{1}{2}$, so we have $\frac{1}{2} - a_n > \frac{1}{2} - \frac{1}{n} \geqslant 0$. Hence, $a_{n+1} < \frac{1}{4} - \left(\frac{1}{2} - \frac{1}{n}\right)^2 = \frac{1}{n} - \frac{1}{n^2} = \frac{n-1}{n^2} < \frac{1}{n+1}$, i.e., the proposition holds true for $n+1$. The proposition is proved.

4. When $n=2$, the proposition is obviously true. Suppose the proposition is true for $n(\geqslant 2)$, then we consider the case $n+1$. From the induction hypothesis, we know that

$$a_1a_2^4 + a_2a_3^4 + \cdots + a_na_{n+1}^4 + a_{n+1}a_1^4$$
$$\geqslant a_2a_1^4 + a_3a_2^4 + \cdots + a_na_{n-1}^4 + a_1a_n^4 - a_na_1^4 + a_na_{n+1}^4 + a_{n+1}a_1^4.$$

In order to prove that the proposition holds true for $n+1$, we only need to prove that

$$a_1a_n^4 - a_na_1^4 + a_na_{n+1}^4 + a_{n+1}a_1^4 \geqslant a_{n+1}a_n^4 + a_1a_{n+1}^4. \quad ①$$

For the sake of convenience, denote $a_1 = x$, $a_n = y$, $a_{n+1} = z$, then $x < y < z$, and to prove ① true is equivalent to having to prove

$$xy^4 + yz^4 + zx^4 - yx^4 - zy^4 - xz^4 \geqslant 0. \quad ②$$

Notice that, the Left Hand Side of ②

$$= xy(y^3 - x^3) + yz(z^3 - y^3) - zx(z^3 - x^3)$$
$$= (xy - zx)(y^3 - x^3) + (yz - zx)(z^3 - y^3)$$
$$= -x(z-y)(y-x)(y^2 + xy + x^2) + z(y-x)(z-y)(z^2 + zy + y^2)$$

$= (y - x)(z - y)(z^3 + z^2y + zy^2 - xy^2 - x^2y - x^3)$

$= (y - x)(z - y)(z - x)(z^2 + zx + x^2 + zy + xy + y^2)$

$= \frac{1}{2}(y - x)(z - y)(z - x)((x + y)^2 + (y + z)^2 + (z + x)^2)$

$\geqslant 0.$

So ② holds true. Furthermore, ① is true. The proposition is true for $n + 1$ and is proved.

5. By condition, we know that $a_{n+1} = a_n + \frac{1}{a_{n-1}}$. Combining the initial conditions and mathematical induction we may know that for any $n \in \mathbf{N}^*$, we have $a_n > 0$. Therefore, for $n \geqslant 2$, we have $a_{n+1} = a_n + \frac{1}{a_{n-1}} > a_n$. Considering together with $a_1 < a_2$ we know that for any $n \in \mathbf{N}^*$, it is true that $a_n < a_{n+1}$. So, when $n \geqslant 2$, we have

$$a_{n+1} = a_n + \frac{1}{a_{n-1}} > a_n + \frac{1}{a_n}. \quad ①$$

From $a_3 = \frac{a_2 a_1 + 1}{a_1} = 3$, we know that $a_3 > \sqrt{6}$, i.e., when $n = 3$, we have $a_n > \sqrt{2n}$. Now we suppose when $n = m(\geqslant 3)$, it is true $a_m > \sqrt{2m}$. Then from ① we know that $a_{m+1}^2 > \left(a_m + \frac{1}{a_m}\right)^2 = a_m^2 + 2 + \frac{1}{a_m^2} > a_m^2 + 2 > 2m + 2$, so $a_{m+1} > \sqrt{2(m + 1)}$. Therefore, the proposition holds true for $m + 1$ and hence is proved.

6. When $n = 1$, from $\frac{1 + a^2}{a} = \frac{1}{a} + a \geqslant 2\sqrt{\frac{1}{a} \cdot a} = 2$ we know that the proposition is true. Now we suppose the proposition is true for n, i.e., $\frac{1 + a^2 + \cdots + a^{2n}}{a + a^3 + \cdots + a^{2n-1}} \geqslant \frac{n + 1}{n}$, then we have

$$\frac{a + a^3 + \cdots + a^{2n-1}}{1 + a^2 + \cdots + a^{2n}} \leqslant \frac{n}{n + 1}.$$

We notice that

$$\frac{1+a^2+\cdots+a^{2n+2}}{a+a^3+\cdots+a^{2n+1}}+\frac{a+a^3+\cdots+a^{2n-1}}{1+a^2+\cdots+a^{2n}}$$

$$=\frac{1+a^2+\cdots+a^{2n+2}}{a(1+a^2+\cdots+a^{2n})}+\frac{a+a^3+\cdots+a^{2n-1}}{1+a^2+\cdots+a^{2n}}$$

$$=\frac{1+a^2+\cdots+a^{2n+2}+a(a+a^3+\cdots+a^{2n-1})}{a(1+a^2+\cdots+a^{2n})}$$

$$=\frac{(1+a^2+\cdots+a^{2n})+a^2(1+a^2+\cdots+a^{2n})}{a(1+a^2+\cdots+a^{2n})}$$

$$=\frac{a^2+1}{a}=a+\frac{1}{a}\geqslant 2.$$

So,

$$\frac{1+a^2+\cdots+a^{2n+2}}{a+a^3+\cdots+a^{2n+1}}\geqslant 2-\frac{n}{n+1}=\frac{n+2}{n+1}.$$

Namely, the proposition is true for $n+1$. QED.

7. When $n=2$, since $2^{10}>1000$, we have $\lg 2>\frac{3}{10}$, the proposition is true. Now we suppose the proposition is true for $n(n\geqslant 2)$. By the Arithmetic Mean-Geometric Mean Inequality, we have $\frac{1+2+\cdots+n}{n}>\sqrt[n]{1\times 2\times\cdots\times n}$, i.e., $n+1>2(n!)^{\frac{1}{n}}$. Then,

$$\begin{aligned}\lg((n+1)!)&>\lg((n!)\cdot 2(n!)^{\frac{1}{n}})\\&=\lg 2+\frac{n+1}{n}\lg(n!)\\&>\lg 2+\frac{n+1}{n}\times\frac{3n}{10}\left(\frac{1}{2}+\cdots+\frac{1}{n}\right)\\&>\frac{3}{10}+\frac{3(n+1)}{10}\left(\frac{1}{2}+\cdots+\frac{1}{n}\right)\\&=\frac{3(n+1)}{10}\left(\frac{1}{2}+\cdots+\frac{1}{n+1}\right).\end{aligned}$$

So the proposition is true for $n+1$. QED.

8. When $n=1$, $a_1^3=a_1^2$, while $a_1>0$, so $a_1=1$, i.e., the proposition is true for $n=1$. Now we suppose the proposition is true for $1, 2, \cdots, n-1$, i.e., $a_k=k$, $k=1, 2, \cdots, n-1$. Then,

$$\left(\sum_{k=1}^{n-1} k^3\right)+a_n^3=\sum_{k=1}^{n} a_k^3=\left(\sum_{k=1}^{n} a_k\right)^2=\left(\left(\sum_{k=1}^{n-1} k\right)+a_n\right)^2.$$

Hence, $a_n^3=a_n^2+n(n-1)a_n$, and we solve it to get $a_n=0$, $-(n-1)$ or n. Consider also $a_n>0$, we have $a_n=n$. So, the proposition is true for n. QED.

9. Without loss of generality, we may set $a_1<a_2<\cdots<a_n$. When $n=1$, the inequality $a_1^2 \geqslant \frac{2+1}{3}a_1$ is true. Assume that the inequality is true for n, i.e., $a_1^2+\cdots+a_n^2 \geqslant \frac{2n+1}{3}(a_1+\cdots+a_n)$, and we consider the case with $n+1$. We only need to prove

$$a_{n+1}^2 \geqslant \frac{2}{3}(a_1+\cdots+a_n)+\frac{2n+3}{3}a_{n+1},$$

where $a_1<a_2<\cdots<a_n<a_{n+1}$, and $a_i \in \mathbf{N}^*$.

Noting that $a_n \leqslant a_{n+1}-1$, $a_{n-1} \leqslant a_n-1 \leqslant a_{n+1}-2$, $\cdots$, $a_1 \leqslant a_{n+1}-n$, so we only need to prove that $a_{n+1}^2 \geqslant \frac{2}{3}\sum_{k=1}^{n}(a_{n+1}-k)+\frac{2n+3}{3}a_{n+1}$, which is equivalent to $a_{n+1}^2-\frac{4n+3}{3}a_{n+1}+\frac{n(n+1)}{3} \geqslant 0$, i.e., we only need to prove that $(a_{n+1}-(n+1))\left(a_{n+1}-\frac{n}{3}\right) \geqslant 0$. This inequality could be proved by making use of $a_{n+1} \geqslant a_1+n \geqslant n+1$.

So, the original inequality is true for $n+1$. QED.

10. From the recursive formula, we can know that $a_{n+1}^2=a_{n-1}a_{n+2}$, $n=2,3,\cdots$. From the initial conditions with mathematical induction, we can know that $a_n \neq 0$. Then, we can turn the above into

$$\frac{a_{n+2}}{a_{n+1}a_n}=\frac{a_{n+1}}{a_n a_{n-1}},\ n=2,3,\cdots.$$

Using this recursively backward, we can know that $\frac{a_{n+2}}{a_{n+1}a_n}=\frac{a_{n+1}}{a_n a_{n-1}}=\cdots=\frac{a_3}{a_2 a_1}=2$, i.e., $a_{n+2}=2a_{n+1}a_n$, $n=2,3,\cdots$. From the recursive expression and that $a_1, a_2, a_3 \in \mathbf{Z}$, we can know that a_n are all integers, and $\frac{a_{n+2}}{a_{n+1}} = 2a_n$ (Note that this equation also holds for $n = 1$.). We may know that for any $n \in \mathbf{N}^*$, $\frac{a_{n+2}}{a_{n+1}}$ is an even number. Then, $a_n = \left(\frac{a_n}{a_{n-1}}\right) \cdot \left(\frac{a_{n-1}}{a_{n-2}}\right) \cdots \left(\frac{a_2}{a_1}\right) \cdot a_1$ and is the product of n even numbers. Hence, $2^n \mid a_n$.

11. From the recursive formula, we may know that $a_{n+1} - k = a_n(a_n - k)$, then $a_n - k = a_{n-1}(a_{n-1} - k) = a_{n-1}a_{n-2}(a_{n-2} - k) = \cdots = a_{n-1} \cdots a_1(a_1 - k) = a_{n-1} \cdots a_1$, i.e.,

$$a_n = a_{n-1} \cdots a_1 + k. \quad ①$$

For any $m, n \in \mathbf{N}^*$, $m \neq n$, without loss of generality, assume $m < n$, and we know from ① that

$$(a_n, a_m) = (a_{n-1} \cdots a_1 + k, a_m) = (k, a_m).$$

Next, we prove: for any $m \in \mathbf{N}^*$, we always have $a_m \equiv 1 \pmod{k}$.

When $m = 1$, from $a_1 = k + 1$ we know that the conclusion is correct. Now we suppose it is correct for m, i.e., $a_m \equiv 1 \pmod{k}$, then $a_{m+1} = a_m^2 - ka_m + k \equiv a_m^2 \equiv 1^2 \equiv 1 \pmod{k}$. So, for any $m \in \mathbf{N}^*$, we have $a_m \equiv 1 \pmod{k}$.

By the above conclusions, we have $(a_m, k) = 1$, and further we get $(a_n, a_m) = 1$.

12. Refer to Example 5 of the first section. Try calculating the first 21 terms of the sequence $\{a_n\}$, and the results are in turn,

$$1, 2, 3, 5, 7, 9, 12, 15, 18, 23, 28, 33,$$
$$40, 47, 54, 63, 72, 81, 93, 105, 117,$$

among which a_1, a_2, a_3, a_4, a_{11}, a_{20} are respectively multiples of 2, 3, 5, 7, 11, 13. Hence, for any $p \in \{2, 3, 5, 7, 11, 13\}$, there is a term a_n, multiple of p.

If $a_{3n-1} \equiv 0 \pmod{p}$, then starting from a_n we can find the next multiple of p. If $a_{3n-1} \not\equiv 0 \pmod{p}$, then from the recursive formula and $a_n \equiv 0 \pmod{p}$, we may know that $a_{3n+2} \equiv a_{3n+1} \equiv a_{3n} \equiv a_{3n-1} \pmod{p}$.

Denote the remainder to be r when they are divided by p. Discuss as the example mentioned before, the 13 numbers below

$$a_{9n-4},\ a_{9n-3},\ \cdots,\ a_{9n+8}$$

are congruent numbers with a_{9n-4}, $a_{9n-4}+r$, $\cdots$, $a_{9n-4}+12r$ under mod p. Since $p \leqslant 13$, these 13 numbers cover at least a complete system of residues for mod p. Then, starting from a_n, we may find the next multiple of p. Proposition proved.

13. Induct with respect to n. Note that

$$\begin{aligned}\sum_{k=1}^{(n+1)^2}\{\sqrt{k}\} &= \sum_{k=1}^{n^2}\{\sqrt{k}\} + \sum_{k=n^2+1}^{(n+1)^2}\{\sqrt{k}\}\\ &\leqslant \frac{1}{2}(n^2-1) + \sum_{k=1}^{2n}(\sqrt{n^2+k}-n)\\ &\leqslant \frac{1}{2}(n^2-1) + \sum_{k=1}^{2n}\left(\sqrt{\left(n+\frac{k}{2n}\right)^2}-n\right)\\ &= \frac{1}{2}(n^2-1) + \frac{1}{2n}\sum_{k=1}^{2n}k\\ &= \frac{1}{2}(n^2-1) + \frac{1}{2}(2n+1)\\ &= \frac{1}{2}((n+1)^2-1).\end{aligned}$$

Then we can show the inductive step.

14. When $n \leqslant m$, inducting with respect to positive integer k, it is easy to prove $k^4 \leqslant 2^{k^2}$, then $\sqrt[k^2]{k^m} \leqslant \sqrt[4]{2^m}$. At this time,

$$\begin{aligned}S_m(n) &\leqslant \sum_{k=1}^{n} k^{\frac{m}{k^2}} = n + \sum_{k=1}^{n}(k^{\frac{m}{k^2}}-1) \leqslant n + \sum_{k=1}^{n}(k^{\frac{m}{k^2}}-1)\\ &\leqslant n + m(2^{\frac{m}{4}}-1),\end{aligned}$$

and the original inequality holds.

When $n > m$, we notice that for any $k \in \mathbf{N}^*$, $k > m$, it is always true that $1 < k^{\frac{m}{k^2}} < k^{\frac{1}{k}} < 2$.

Here $k^{\frac{1}{k}} < 2$ is equivalent to $k < 2^k$ and can be proved through inducting with respect to k. Then $S_m(n+1) = S_m(n)+1$. Considering how the inequality holds for $n \leqslant m$ and mathematical induction, we may know that for any m, $n \in \mathbf{N}^*$, it is always true that $S_m(n) \leqslant n + m(\sqrt[4]{2^m} - 1)$.

15. We use A_k to represent the sets whose elements are the integers from the sequence $\{\sqrt[k]{a_n}\}$. We assert that for any $k \in \mathbf{N}^*$, $A_k = \{2^m \mid m = 0, 1, 2, \cdots\}$. (i.e., A_k is irrelevant of the concrete value of k, and is formed by powers of 2.)

From $a_0 = 1$, we know that $1 \in A_k$. Next, we suppose $x \in A_k$. We will prove that among all the numbers in A_k that are bigger than x, the smallest one is $2x$. According to this conclusion and by mathematical induction, we may know that our earlier assertion is correct.

As a matter of fact, suppose $x \in A_k$, i.e., there exists an $n \in \mathbf{N}$, such that $a_n = x^k$, then for any subscript j that satisfies $x^k \leqslant a_j < (x+1)^k$, we have $a_{j+1} = a_j + x$, i.e., $a_{j+1} \equiv a_j \pmod{x}$. From a_n, we can know that for this kind of j, it is true that $a_{j+1} \equiv a_j \equiv 0 \pmod{x}$. Now we take the biggest j that meets the above conditions, then at this time, $a_j < (x+1)^k$, while $a_{j+1} \geqslant (x+1)^k$. Denote $a_{j+1} = (x+1)^k + m_1$, then from $a_{j+1} = a_j + x$ we know that $0 \leqslant m_1 < x$, and because $a_{j+1} \equiv 0 \pmod{x}$, it is true that $m_1 + 1 \equiv 0 \pmod{x}$. So $m_1 = x - 1$. Subsequently, $a_{j+1} = (x+1)^k + x - 1$.

Repeating the above discussions, by adding $x+1$ every time, we get the terms in the form of $(x+2)^k + m_2$, where $0 \leqslant m_2 < n+1$. From $x - 1 \equiv (x+2)^k + m_2 \equiv 1 + m_2 \pmod{x+1}$, we may ascertain $m_2 = x - 2$. Conducting the above process recursively, we ascertain that $m_i = x - i$, $i = 1, 2, \cdots, x$ by making use of the congruence $m_i \equiv (x+i+1)^k + m_{i+1} \pmod{x+i}$, in general form.

Hence the next kth power will occur in the sequence $\{a_n\}$, when m_i for the first time takes on the value of zero while $i = x$, i.e., the next kth power number is $(x+x)^k = (2x)^k$. That is to say, among the numbers in A_k that are bigger than x, the smallest one is $2x$. The problem is solved.

16. From the recursive relationship, we know that for any $n \in \mathbf{N}^*$, $x_n > 0$. Furthermore, from $2nx_n = 2(n-1)x_{n-1} - x_{n-1}$ we can get $x_{n-1} = 2(n-1)x_{n-1} - 2nx_n$. Taking the sum, and we get

$$\begin{aligned} x_1 + \cdots + x_n &= \sum_{k=2}^{n+1}(2(k-1)x_{k-1} - 2kx_k) \\ &= 2\sum_{k=1}^{n}(kx_k - (k+1)x_{k+1}) \\ &= 2(x_1 - (n+1)x_{n+1}) \\ &= 1 - 2(n+1)x_{n+1} < 1. \end{aligned}$$

The proposition is proved.

17. From the conditions, we know that $f(n+1) = (f(n)-1)f(n) + 1$. Thinking about mathematical induction and $a_1 > 1$, we can get for any $n \in \mathbf{N}^*$, it is always true that $f(n) > 1$. Then, by taking the reciprocal, we have

$$\frac{1}{f(n+1)-1} = \frac{1}{f(n)(f(n)-1)} = \frac{1}{f(n)-1} - \frac{1}{f(n)}.$$

i.e., $\dfrac{1}{f(n)} = \dfrac{1}{f(n)-1} - \dfrac{1}{f(n+1)-1}$. By telescoping series, we have

$$\sum_{k=1}^{n}\frac{1}{f(k)} = \frac{1}{f(1)-1} - \frac{1}{f(n+1)-1} = 1 - \frac{1}{f(n+1)-1}.$$

Coming back to the recursive formula, we have

$$\begin{aligned} f(n+1) - 1 &= f(n)(f(n)-1) > (f(n)-1)^2 > (f(n-1)-1)^{2^2} \\ &> \cdots > (f(2)-1)^{2^{n-1}} = (2^2-2)^{2^{n-1}} = 2^{2^{n-1}}. \end{aligned}$$

So, $\displaystyle\sum_{k=1}^{n}\frac{1}{f(k)} > 1 - \frac{1}{2^{2^{n-1}}}$.

On the other hand, $f(n+1) = f(n)^2 - (f(n)-1) < f(n)^2$. So,

$$f(n+1) < f(n)^2 < f(n-1)^{2^2} < \cdots < f(1)^{2^n} = 2^{2^n},$$

and $f(n+1) - 1 < 2^{2^n}$. Further, $\displaystyle\sum_{k=1}^{n}\frac{1}{f(k)} < 1 - \frac{1}{2^{2^n}}$. The proposition is proved.

18. Denote $x_1 = \cot\alpha$, $y_1 = \tan\beta$, here $\alpha = \frac{\pi}{6}$, $\beta = \frac{\pi}{3}$. Then,

$$x_2 = \cot\alpha + \csc\alpha = \frac{1+\cos\alpha}{\sin\alpha} = \frac{2\cos^2\frac{\alpha}{2}}{2\sin\frac{\alpha}{2}\cos\frac{\alpha}{2}} = \cot\frac{\alpha}{2}.$$

According to this and mathematical induction, it is easy to prove that $x_n = \cot\frac{\alpha}{2^{n-1}}$. Similarly, we may prove $y_n = \tan\frac{\beta}{2^{n-1}}$. Therefore, when $n > 1$, we have

$$x_n y_n = \cot\frac{\alpha}{2^{n-1}}\tan\frac{\beta}{2^{n-1}} = \cot\frac{\pi}{2^n \times 3}\tan\frac{\pi}{2^{n-1}\times 3} = \frac{2}{1-\tan^2\frac{\pi}{2^n\times 3}}.$$

Since $\tan^2\frac{\pi}{2^n\times 3} \in \left(0, \tan^2\frac{\pi}{6}\right)$, namely, $\tan^2\frac{\pi}{2^n\times 3} \in \left(0, \frac{1}{3}\right)$, it is true that $2 < x_n y_n < 3$. The proposition is proved.

19. From the given conditions, we may know that $c_n - 1 = (c_{n-1} - 1)^2$, then $c_n - 1 = (c_{n-1} - 1)^2 = (c_{n-2} - 1)^4 = \cdots = (c_0 - 1)^{2^n} = 3^{2^n}$, so $c_n = 3^{2^n} + 1$.

On the other hand, $1 - a_{n+1} = \frac{(1-a_n)^2}{1+a_n^2}$, $1 + a_{n+1} = \frac{(1+a_n)^2}{1+a_n^2}$, then $\frac{1-a_{n+1}}{a+a_{n+1}} = \left(\frac{1-a_n}{1+a_n}\right)^2$, moreover,

$$\frac{1-a_n}{1+a_n} = \left(\frac{1-a_{n-1}}{1+a_{n-1}}\right)^2 = \cdots = \left(\frac{1-a_0}{1+a_0}\right)^{2^n} = \left(\frac{1}{3}\right)^{2^n},$$

so $a_n = \frac{3^{2^n}-1}{3^{2^n}+1}$.

Noting that

$$\begin{aligned} 2c_0c_1 \cdot \cdots \cdot c_{n-1} &= (3-1)(3+1)(3^2+1)\cdot \cdots \cdot (3^{2^{n-1}}+1) \\ &= (3^2-1)(3^2+1)\cdot \cdots \cdot (3^{2^{n-1}}+1) \\ &= \cdots = 3^{2^n} - 1. \end{aligned}$$

So the proposition is true.

20. Denote $f(x)=\frac{x}{n}+\frac{n}{x}$, then from $f(a)-f(b)=\frac{(a-b)(ab-n^2)}{abn}$, we know that the function $f(x)$ is a decreasing function on the interval $(0, n]$.

Next, we are going to use mathematical induction with respect to n. Firstly we are to prove $\sqrt{n}<a_n<\frac{n}{\sqrt{n-1}}$, $n\geqslant 3$. Noting that $a_1=1$, we know that $a_2=2$, $a_3=2$. Then when $n=3$, the above inequality is true. Furthermore, suppose $\sqrt{n}<a_n<\frac{n}{\sqrt{n-1}}$, $n\geqslant 3$. From monotonicity we know that $f(a_n)<f(\sqrt{n})=\frac{n+1}{\sqrt{n}}$, i. e., $a_{n+1}<\frac{n+1}{\sqrt{n}}$, and

$$a_{n+1}=f(a_n)>f\left(\frac{n}{\sqrt{n-1}}\right)=\frac{n}{\sqrt{n-1}}>\sqrt{n+1}.$$

So for any $n\in\mathbf{N}^*$, $n\geqslant 3$, it is always true that

$$\sqrt{n}<a_n<\frac{n}{\sqrt{n-1}}.$$

Next, we are to prove that when $n\geqslant 4$, $a_n<\sqrt{n+1}$.

As a matter of fact, since when $n\geqslant 3$, $a_{n+1}=f(a_n)>f\left(\frac{n}{\sqrt{n-1}}\right)=\frac{n}{\sqrt{n-1}}$. So when $n\geqslant 4$, we have $a_n>\frac{n-1}{\sqrt{n-2}}$. Further, when $n\geqslant 4$, we have

$$a_{n+1}=f(a_n)<f\left(\frac{n-1}{\sqrt{n-2}}\right)=\frac{(n-1)^2+n^2(n-2)}{(n-1)n\sqrt{n-2}}<\sqrt{n+2}.$$

(The last inequality is equivalent to $2n^2(n-3)+4n-1>0$.) And it is evident that $a_4=\frac{13}{6}<\sqrt{6}$.

Then, when $n\geqslant 4$, it is always true that $\sqrt{n}<a_n<\sqrt{n+1}$.

Hence, we have $[a_n^2] = n$.

21. If there exists an $n \in \mathbf{N}^*$, such that $(a+\sqrt{a^2-1})^{\frac{1}{n}} + (a-\sqrt{a^2-1})^{\frac{1}{n}}$ is a rational number, while denoting $x = (a+\sqrt{a^2-1})^{\frac{1}{n}}$, $y = (a-\sqrt{a^2-1})^{\frac{1}{n}}$, then $x+y$ is a rational number, and $x^n + y^n = 2a$ is an irrational number.

We will prove this by inducting with respect to m. If $x+y \in \mathbf{Q}$, then for any $m \in \mathbf{N}^*$, it is always true that $x^m + y^m \in \mathbf{Q}$. ①

Noting that $x^2+y^2 = (x+y)^2 - 2xy = (x+y)^2 - 2$, together with $x+y \in \mathbf{Q}$, we may know that when $m = 1, 2$, ① is always true. Now suppose $x^m + y^m$, $x^{m+1} + y^{m+1} \in \mathbf{Q}$. Then from $x^{m+2} + y^{m+2} = (x+y)(x^{m+1}+y^{m+1}) - xy(x^m+y^m)$, and the fact that both $x+y$, $xy(=1)$ are rational numbers, we know that $x^{m+2}+y^{m+2} \in \mathbf{Q}$. Then ① is true.

By ① we know that $x^n + y^n \in \mathbf{Q}$, and this is a contradiction. So the proposition is true.

22. If $t>1$, then $a_2<0$. By this and mathematical induction, we may know that for $n \geqslant 2$, it is always true that $a_n < 0$, so $a_{2011} \neq 0$; If $t<0$, by the same reasoning as above we can know that when $n \geqslant 1$, it is always true that $a_n < 0$, and there is no chance that $a_{2011} = 0$. Hence, the t that makes $a_{2011} = 0$ satisfies $t \in [0, 1]$.

Now we may suppose $t = \sin^2\alpha$, where $0 \leqslant \alpha \leqslant \frac{\pi}{2}$, then $a_1 = \sin^2\alpha$. If $a_n = \sin^2(2^{n-1}\alpha)$, then $a_{n+1} = 4\sin^2(2^{n-1}\alpha)\cos^2(2^{n-1}\alpha) = \sin^2(2^n\alpha)$. Then, from principles of mathematical induction we know that for any n, we have $a_n = \sin^2(2^{n-1}\alpha)$. Hence, from $a_{2011} = 0$, we get $\sin^2(2^{2010}\alpha) = 0$, and hence $2^{2010}\alpha = k\pi$, i.e., $\alpha = \frac{k\pi}{2^{2010}}$, $k \in \mathbf{Z}$. Combining $0 \leqslant \alpha \leqslant \frac{\pi}{2}$, we know that $0 \leqslant k \leqslant 2^{2009}$. Noting that the sine function is non-negative on the interval $\left[0, \frac{\pi}{2}\right]$ and is monotonically increasing, we know that there exist $2^{2009}+1$ distinct real numbers t, such that $a_{2011} = 0$.

23. The maximum is $(1+2+\cdots+1005)\times 2 = 1\,011\,030$. It occurs when $x_1 = 1005$, $x_2 = 1004$, $\cdots$, $x_{1005} = 1$, $x_{1006} = 0$, $x_{1007} = -1$, $\cdots$,

$x_{2011}=-1005$.

Now we will prove that for a sequence that meets the conditions, it is true that

$$\sum_{i=1}^{2011}|x_i|-\left|\sum_{i=1}^{2011}x_i\right|\leqslant 2(1+2+\cdots+1005). \quad ①$$

Notice that after we rearrange x_1, $\cdots$, x_{2011}, from biggest to smallest, as y_1, y_2, $\cdots$, y_{2011}, for $1\leqslant i\leqslant 2010$, suppose $y_i=x_m$, $y_{i+1}=x_n$. We may always find a subscript j, such that $x_j\in\{y_1,\cdots,y_i\}$, $x_{j+1}\in\{y_{i+1},\cdots,y_{2011}\}$, or $x_j\in\{y_{i+1},\cdots,y_{2011}\}$, $x_{j+1}\in\{y_1,\cdots,y_i\}$. (This conclusion may be derived through proof by contradiction, combining the two cases $x_1\in\{y_1,\cdots,y_i\}$ and $x_1\in\{y_{i+1},\cdots,y_{2011}\}$). Without loss of generality, we suppose the former one is true, and also suppose that $x_j=y_r$, $x_{j+1}=y_t$, then $r\leqslant i$, $t\geqslant i+1$. At this time,

$$\begin{aligned}1\geqslant|x_j-x_{j+1}|&=|y_r-y_t|\\&=|(y_r-y_{r+1})+(y_{r+1}-y_{r+2})+\cdots+(y_i-y_{i+1})+\cdots+(y_{t-1}-y_t)|\\&=|y_r-y_{r+1}|+|y_{r+1}-y_{r+2}|+\cdots+|y_i-y_{i+1}|+\cdots+|y_{t-1}-y_t|\\&\geqslant|y_i-y_{i+1}|.\end{aligned}$$

(Here we used the decreasing permutation of y_1, $\cdots$, y_{2011}.)

Hence, we still have $|y_i-y_{i+1}|\leqslant 1$.

Furthermore, without loss of generality, we suppose $\sum_{i=1}^{2011}x_i\leqslant 0$. $\left(\text{If } \sum_{i=1}^{2011}x_i>0, \text{ then we replace } x_i \text{ by } -x_i \text{ and then proceed to discussions.}\right)$ After the re-ordering, suppose $y_1\geqslant\cdots\geqslant y_k\geqslant 0\geqslant y_{k+1}\geqslant\cdots\geqslant y_{2011}$, then

$$\begin{aligned}S&=\sum_{i=1}^{2011}|x_i|-\left|\sum_{i=1}^{2011}x_i\right|\\&=(y_1+\cdots+y_k)-(y_{k+1}+\cdots+y_{2011})+(y_1+\cdots+y_{2011})\\&=2(y_1+\cdots+y_k).\end{aligned}$$

In order to prove that ① is true, we only need to prove that

$$y_1 + \cdots + y_k \leqslant 1 + 2 + \cdots + 1005. \quad ②$$

We deal with this with two cases.

Case 1. If $k \geqslant 1006$, then from $y_1 + \cdots + y_{2011} \leqslant 0$, we know that

$$y_1 + \cdots + y_k \leqslant -(y_{k+1} + \cdots + y_{2011}).$$

Noting also $y_{k+1} \geqslant y_k - 1$, $\cdots$, $y_{2011} \geqslant y_k - (2011 - k)$, we know that

$$\begin{aligned} y_1 + \cdots + y_k &\leqslant -((y_k - 1) + \cdots + (y_k - (2011 - k))) \\ &= -(2011 - k)y_k + 1 + 2 + \cdots + (2011 - k) \\ &\leqslant 1 + 2 + \cdots + (2011 - k) \\ &\leqslant 1 + 2 + \cdots + 1005. \end{aligned}$$

Thus, ② holds true.

Case 2. If $k \leqslant 1005$, then by similar reasoning as above, we may know that

$$\begin{aligned} y_1 + \cdots + y_k &\leqslant (y_{k+1} + k) + (y_{k+1} + (k - 1)) + \cdots + (y_{k+1} + 1) \\ &= k y_{k+1} + 1 + \cdots + k \leqslant 1 + 2 + \cdots + k \\ &\leqslant 1 + 2 + \cdots + 1005. \end{aligned}$$

Thus, ② also holds true.

By all above, the maximum value is 1 011 030.

24. We use proof by contradiction. If the proposition is not true, then, there exists a positive integer N, such that for any $n \geqslant N$, it is always true that

$$1 + a_n \leqslant \sqrt[n]{2} \cdot a_{n-1}. \quad ①$$

Now we define a sequence $\{c_n\}$ of positive real numbers:

$$c_0 = 1, \; c_n = \frac{a_{n-1}}{1 + a_n} c_{n-1}, \; n = 1, 2, \cdots.$$

Then from ① we may know that for any $n \geqslant N$, it is always true that

$$c_n \geqslant 2^{-\frac{1}{n}} \cdot c_{n-1}. \quad ②$$

Notice that for $n \in \mathbf{N}^*$, it is true $c_n(1+a_n) = a_{n-1}c_{n-1}$, i.e., $c_n = a_{n-1}c_{n-1} - a_nc_n$. By using telescoping series, we get

$$c_1 + \cdots + c_n = a_0 - a_nc_n < a_0.$$

This suggests that the nth partial sum sequence $\{s_n\}$, where $s_n = c_1 + \cdots + c_n$, is bounded.

On the other hand, from ② we can know that when $n > N$, it is true that

$$\begin{aligned} c_n &\geqslant c_{n-1} \cdot 2^{-\frac{1}{n}} \geqslant c_{n-2} \cdot 2^{-\left(\frac{1}{n-1}+\frac{1}{n}\right)} \\ &\geqslant \cdots \geqslant c_N \cdot 2^{-\left(\frac{1}{N+1}+\cdots+\frac{1}{n}\right)} \\ &= C \cdot 2^{-\left(1+\frac{1}{2}+\cdots+\frac{1}{n}\right)}. \end{aligned}$$

Here $C = c_N \cdot 2^{-\left(1+\frac{1}{2}+\cdots+\frac{1}{N}\right)}$ is a constant.

For any $k \in \mathbf{N}^*$, if $2^{k-1} \leqslant n < 2^k$, then

$$\begin{aligned} & 1 + \frac{1}{2} + \cdots + \frac{1}{n} \\ \leqslant\ & 1 + \left(\frac{1}{2} + \frac{1}{3}\right) + \left(\frac{1}{4} + \cdots + \frac{1}{7}\right) + \cdots + \left(\frac{1}{2^{k-1}} + \cdots + \frac{1}{2^k - 1}\right) \\ \leqslant\ & 1 + \left(\frac{1}{2} + \frac{1}{2}\right) + \left(\frac{1}{4} + \cdots + \frac{1}{4}\right) + \cdots + \left(\frac{1}{2^{k-1}} + \cdots + \frac{1}{2^{k-1}}\right) = k. \end{aligned}$$

So, at this time, we have $c_n \geqslant C \cdot 2^{-k}\ (2^{k-1} \leqslant n < 2^k)$.

Now we suppose $2^{r-1} \leqslant N < 2^r$, $r \in \mathbf{N}^*$, then for any $m > r$, we have

$$\begin{aligned} & c_{2^r} + c_{2^r+1} + \cdots + c_{2^m-1} \\ =\ & (c_{2^r} + \cdots + c_{2^{r+1}-1}) + \cdots + (c_{2^{m-1}} + \cdots + c_{2^m-1}) \\ \geqslant\ & (C \cdot 2^{-(r+1)}) \cdot 2^r + \cdots + (C \cdot 2^{-(m+1)}) \cdot 2^m \\ =\ & \frac{C(m-r)}{2}. \end{aligned}$$

This suggests that $s_{2^m-1} > \frac{C(m-r)}{2}$. When $m \to +\infty$, we have $s_{2^m-1} \to +\infty$, contradictory to the conclusion that the sequence $\{s_n\}$ is bounded.

So, the proposition is true.

25. In condition (1), let $n=0$, and we know that $F(0)=0$. For $n \in \mathbf{N}^*$, suppose the binary expression for n is $n=(n_k n_{k-1} \cdots n_0)=n_k \cdot 2^k+\cdots+n_0 \cdot 2^0$, where $n_k=1$. And for $0 \leqslant i \leqslant k-1$, we have $n_i \in \{0, 1\}$.

Now we prove through inducting with respect to k: for any $n \in \mathbf{N}^*$, it is always true that

$$F(n)=n_k F_k+n_{k-1} F_{k-1}+\cdots+n_0 F_0, \qquad ①$$

Here the sequence $\{F_m\}$ is defined as $F_0=F_1=1$, $F_{m+2}=F_{m+1}+F_m$, $m=0, 1, 2, \cdots$ (It is defined by translating each subscript for a Fibonacci Sequence forward by one unit.)

As a matter of fact, from $F(0)=0$ and condition (3) we may know that $F(1)=1$. Further, we can get $F(2)=1$, $F(3)=F(2)+1=F_0+F_1$, $F(4)=F(2)+F(1)=2=F_2$. So, the proposition is true for $k=0, 1$. Now we suppose that ① is true for k and $k+1$. We consider the case with $k+2$. Now we may suppose $n=(n_{k+2} n_{k+1} \cdots n_0)_2$. If $(n_1, n_0)=(0, 0)$, then from (1) we know that $F(n)=F((n_{k+2} n_{k+1} \cdots n_1)_2)+F((n_{k+2} \cdots n_2)_2)=n_{k+2} F_{k+1}+\cdots+n_1 F_0+n_{k+2} F_k+\cdots+n_2 F_0=n_{k+2}(F_{k+1}+F_k)+\cdots+n_2(F_1+F_0)+n_1 F_0=n_{k+2} F_{k+2}+\cdots+n_2 F_2+n_1 F_1+n_0 F_0$ (here we make use of $n_1=n_0=0$), ① is true for $k+2$; If $(n_1, n_0)=(1, 0)$, then from (2) we know that $F(n)=F((n_{k+2} n_{k+1} \cdots n_2 n_1' n_0')_2)+1$, where $n_1'=n_0'=0$. Then,

$$\begin{aligned} F(n) &= n_{k+2} F_{k+2}+\cdots+n_2 F_2+1 \\ &= n_{k+2} F_{k+2}+\cdots+n_2 F_2+n_1 F_1+n_0 F_0, \end{aligned} \qquad ①$$

also holds true; if $(n_1, n_0)=(0, 1)$, then $F(n)=F((n_{k+2} \cdots n_2 n_1' n_0')_2)+1$, where $n_1'=n_0'=0$. By condition (1) and previous conclusions we know that ① holds true; if $(n_1, n_0)=(1, 1)$, then $F(n)=F((n_{k+2} \cdots n_1 n_0'))+1$, where $n_0'=0$. By condition (2) and previous conclusions we know that ① holds true. So, ① holds true for any $n \in \mathbf{N}^*$.

Making use of ① we may know that the sufficient and necessary condition for $F(4n)=F(3n)$ is that within the binary expression for

n, $(n_k n_{k-1} \cdots n_0)_2$, there are no two adjacent numbers who are both 1 (here we also use the definition of the sequence $\{F_m\}$). For $0 \leqslant n < 2^m$, we denote the number of binary expressions with no adjacent 1s to be f_m, then $f_0 = 1$, $f_1 = 2$. In the meantime, if we delete the last digit n_0 from n, then by classification according to $n_0 = 0$ or 1, we have respectively f_{m-1} and f_{m-2} (since when $n_0 = 1$, it must be true that $n_1 = 0$). So, $f_m = f_{m-1} + f_{m-2}$. This suggests that for $0 \leqslant n < 2^m$, the number n of $F_{m+1} (= F(2^{m+1}))$ satisfies the equation $F(4n) = F(3n)$, so the proposition is true.

26. From the recurrence formula, we may know that

$$f(n) \leqslant f(n-1) + 2 \leqslant \cdots \leqslant f(1) + 2(n-1) = 2n - 1.$$

So, $f(n) - n + 1 \leqslant n$.

Hence, if the values of $f(1)$, $\cdots$, $f(n)$ are determined, then the value of $f(n+1)$ can be uniquely determined. Therefore, there exists a unique function f that meets the conditions. Now, let $g(n) = \left[\frac{1+\sqrt{5}}{2} n\right]$. Denote $\alpha = \frac{1+\sqrt{5}}{2}$, then $g(1) = 1$, and for any $n \in \mathbf{N}^*$, we always have $g(n+1) - g(n) = [\alpha(n+1)] - [\alpha n] = [\alpha + \varepsilon]$, where

$$\varepsilon = \{\alpha n\} = \alpha n - [\alpha n].$$

On the other hand,

$$\begin{aligned} g(g(n) - n + 1) &= [\alpha(g(n) - n + 1)] = [\alpha(\alpha n - \varepsilon - n + 1)] \\ &= [(\alpha^2 - \alpha)n + \alpha(1 - \varepsilon)] = n + [\alpha(1 - \varepsilon)]. \end{aligned}$$

Here we use $\alpha^2 - \alpha - 1 = 0$.

Note that $\varepsilon \neq 2 - \alpha = \frac{3 - \sqrt{5}}{2}$ (otherwise $1 = \frac{[\alpha n] + \varepsilon}{\alpha} = \frac{[\alpha n] + 2}{\alpha} - 1$, leading to a contradictory conclusion that α is a rational number). We make use of the above conclusion and find that if $0 \leqslant \varepsilon < 2 - \alpha$, then $\alpha(1 - \varepsilon) > \alpha(\alpha - 1) = 1$, then $g(g(n) - n + 1) = n + 1$. At this time, $1 < \alpha + \varepsilon < \alpha + 2 - \alpha = 2$, i.e., $g(n+1) - g(n) = 1$; if $2 - \alpha < \varepsilon < 1$, then $\alpha(1 - \varepsilon) < \alpha(\alpha - 1) = 1$, then $g(g(n) - n + 1) = n$. At this time, $2 < \alpha + \varepsilon < 3$, i.e., $g(n+1) - g(n) = 2$.

The above discussions show that $g: \mathbf{N}^* \to \mathbf{N}^*$ meets all the conditions that f meets, and therefore for any $n \in \mathbf{N}^*$, we have $f(n) = g(n)$. This gives the answer (2) requires for.

Combining with ①, we know that (1) holds. QED.

27. Suppose under base 2, for any $n \in \mathbf{N}^*$, the total number of appearances of 00 and 11 among all adjacent number pairs is denoted as x_n, while the total number of appearances of 01 and 10 among all adjacent number pairs is denoted as y_n. We will prove that $a_n = x_n - y_n$. ①

As a matter of fact, when $n = 1$, $x_1 = y_1 = 0$, so ① is true for $n = 1$.

Now we suppose ① is true for the subscripts $1, 2, \cdots, n-1$ $(n \geqslant 2)$. Consider the case n.

If under base 2, the last two digits of n are either 00 or 11, then $n \equiv 0$, or $3 \pmod 4$. At this time, $a_n = a_{\left[\frac{n}{2}\right]} + 1$, while $x_n = x_{\left[\frac{n}{2}\right]} + 1$, $y_n = y_{\left[\frac{n}{2}\right]}$. So ① is true for n.

If under base 2, the last two digits of n are either 01 or 10, then $n \equiv 1$, or $2 \pmod 4$. At this time, $a_n = a_{\left[\frac{n}{2}\right]} - 1$, while $x_n = x_{\left[\frac{n}{2}\right]}$, $y_n = y_{\left[\frac{n}{2}\right]} + 1$. So ① is true for n.

For all above, ① is true for any $n \in \mathbf{N}^*$.

Now we need to calculate among $2^k \leqslant n < 2^{k+1}$, the number of n that makes x_n equal to y_n under base 2.

Note that, under base 2, n is a $k+1$ digit number and let it be B_n. When $k \geqslant 1$, subtract the next digit number, left to right, from every digit number of B_n, and then take absolute values for each digit. We can then get a k-element array C_n comprised of 0 or 1. (For example, if $B_n = (1101)_2$, then $C_n = (011)_2$.) Note that every adjacent number pairs 00 and 11 change into one 0 in C_n, while 01 and 10 change into one 1 in C_n. So, if $x_n = y_n$, then the numbers of 1s and 0s in C_n are the same. Conversely, for a k-element array $C_n = (C_1 C_2 \cdots C_k)$ comprised of 0 or 1, under mod 2 we may find the sums of $b_1 = 1 + c_1$, $b_2 = b_1 + c_2$, $\cdots$, $b_k = b_{k-1} + b_k$, where $b_0 = 1$. Then $B_n = (b_0 b_1 \cdots b_k)_2$

is a binary expression for the number n satisfying $2^k \leqslant n < 2^{k+1}$. This suggests that B_n and C_n have a one-on-one correspondence.

So, the answer to the original question is equal to the number of arrays, which are k-element arrays comprised of 0 or 1, that have equal number of 0s and 1s. Therefore, when k is an odd number, the answer is 0; when k is an even number, the answer is $C_k^{\frac{k}{2}}$. (note: we deem $C_0^0 = 1$.)

28. We are going to prove: for any $n \in \mathbf{N}$, it is always true that

$$x_{5n+1} = 5n+1,\ x_{5n+2} = 5n+4,\ x_{5n+3} = 5n+2,$$
$$x_{5n+4} = 5n+5,\ x_{5n+5} = 5n+3. \quad ①$$

(By making use of this result and $k^2 \equiv 0$, 1, or 4(mod 5), we may know that the proposition holds true.)

As a matter of fact, when $n = 0$, from $a_1 = 1$ we know that

$$a_2 = 4,\ a_3 = 2,\ a_4 = 5,\ a_5 = 3.$$

So ① is true for $n = 0$.

Now we suppose that ① is true for $n = 0, 1, 2, \cdots, m-1$ ($m \in \mathbf{N}^*$). Now consider the case $n = m$. From the structure of ① ($a_{5n+1}, \cdots, a_{5n+5}$ is a permutation of $5n+1, \cdots, 5n+5$), we know that $a_1, a_2, \cdots, a_{5m}$ is a permutation of $1, 2, \cdots, 5m$. By making use of recursive relationship we may know that

$$a_{5m+1} = a_{5m} - 2 = 5m+1,\ a_{5m+2} = a_{5m+1} + 3 = 5m+4,$$
$$a_{5m+3} = a_{5m+2} - 2 = 5m+2,\ a_{5m+4} = a_{5m+3} + 3 = 5m+5,$$
$$a_{5m+5} = a_{5m+4} - 2 = 5m+2.$$

So, the conclusion ① is also true for m.

29. Using that $\dfrac{\pi}{\theta}$ is an irrational number, we know $a_1, a_2, \cdots, a_n$ are n distinctive real numbers. To ascertain the value of the algebraic expression, we will look for a polynomial of degree n with $a_1, \cdots, a_n$ as roots.

Note that $e^{i\theta} = \cos\theta + i\sin\theta, e^{-i\theta} = \cos\theta - i\sin\theta$. Then, we have

$\sec\theta\, e^{i\theta} = 1 + i\tan\theta$, $\sec\theta \cdot e^{-i\theta} = 1 - i\tan\theta$. So,

$$1 + i\tan\theta = e^{2i\theta}(1 - i\tan\theta). \qquad ①$$

Let $\omega = e^{2in\theta}$, then the polynomial $Q_n(x) = (1 + ix)^n - \omega(1 - ix)^n$ has n roots $a_1, a_2, \cdots, a_n$. (This can be known from ①, since the nth roots of ω are $e^{2i(\theta+\frac{k}{n}\pi)}$, $k = 1, 2, \cdots, n$.) Since $Q_n(x)$ is a polynomial of degree n, $a_1, \cdots, a_n$ are all the roots that $Q_n(x)$ have.

Denote $Q_n(x) = c_n x^n + \cdots + c_0$. Then by Vieta's formulas, we may know that $a_1 + \cdots + a_n = -\frac{c_{n-1}}{c_n}$, $a_1 \cdots a_n = (-1)^n \cdot \frac{c_0}{c_n}$. So,

$$\frac{a_1 + \cdots + a_n}{a_1 \cdots a_n} = (-1)^{n-1} \cdot \frac{c_{n-1}}{c_0}.$$

We apply binomial theorem on $Q_n(x)$ and see that

$$c_{n-1} = n \cdot i^{n-1} - \omega n(-i)^{n-1} = ni^{n-1}(1 - \omega), c_0 = 1 - \omega,$$

thus $\frac{a_1 + \cdots + a_n}{a_1 \cdots a_n} = n \cdot (-i)^{n-1}$.

Considering that n is an odd number, we have

$$\frac{a_1 + \cdots + a_n}{a_1 \cdots a_n} = (-1)^{\frac{n-1}{2}} \cdot n.$$

The problem is solved.

30. When $n = 1$, just take $P(x) = x$. When $n = 2$, $2\cos 2\varphi = (2\cos\varphi)^2 - 2$, the proposition also holds true.

Suppose the proposition holds true for $n = k$ and $k + 1$, namely, there exist polynomials, $f(x)$ and $g(x)$, with integral coefficients whose leading coefficients are both 1, such that

$$2\cos k\varphi = f(2\cos\varphi), \ 2\cos(k + 1)\varphi = g(2\cos\varphi).$$

The degrees of f, g are k and $k + 1$, respectively.

Next, we consider the case with $n = k + 2$. Note that

$$\begin{aligned} 2\cos(k + 2)\varphi &= 2\cos[(k + 1)\varphi + \varphi] \\ &= 2\cos(k + 1)\varphi\cos\varphi - 2\sin(k + 1)\varphi\sin\varphi. \qquad ① \end{aligned}$$

$$2\cos k\varphi = 2\cos[(k + 1)\varphi - \varphi]$$

$$=2\cos(k+1)\varphi\cos\varphi+2\sin(k+1)\varphi\cos\varphi. \qquad ②$$

Add up ① and ②, we can get

$$2\cos(k+2)\varphi+2\cos k\varphi=4\cos(k+1)\varphi\cos\varphi.$$

By making use of induction hypothesis, we can know that

$$2\cos(k+2)\varphi=(2\cos\varphi)g(2\cos\varphi)-f(2\cos\varphi).$$

So, we let $h(x)=xg(x)-f(x)$ (it is easy to know that $h(x)$ is a polynomial with integral coefficients whose leading coefficient is 1) and we have $2\cos(k+2)\varphi=h(2\cos\varphi)$.

The proposition is true for $k+2$.

So, the proposition holds true.

31. Denote $\theta=\alpha\pi$. Since α is a rational number, we know that there exists $n\in\mathbf{N}^*$ such that $n\theta=2k\pi$, $k\in\mathbf{Z}$, i.e., $\cos n\theta=1$. From the conclusion of the above problem, we know that there exists a polynomial with integral coefficients $f(x)=x^n+a_{n-1}x^{n-1}+\cdots+a_0$, such that $2\cos n\theta=f(2\cos\theta)$. Therefore,

$$(2\cos\theta)^n+a_{n-1}(2\cos\theta)^{n-1}+\cdots+a_1(2\cos\theta)+a_0-2=0.$$

This suggests that $2\cos\theta$ (attention that $\cos\alpha\pi\in\mathbf{Q}$) is a rational root of the equation below

$$x^n+a_{n-1}x^{n-1}+\cdots+a_1x+a_0-2=0. \qquad ①$$

However, the left-hand side is a polynomial with leading coefficient 1. Therefore, the rational roots of ① are all integers. So, $2\cos\theta$ is an integer. Considering also $|\cos\theta|\leqslant 1$, we know that $2\cos\theta\in\{-2,-1,0,1,2\}$, and then we have $\cos\alpha\pi\in\left\{0,\pm\frac{1}{2},\pm 1\right\}$. (It is obvious that for any value in the set, there is a value of α corresponding to that value.)

32. Without loss of generality, we suppose the equation of the unit circle is $x^2+y^2=1$. Now we take $\theta=\arccos\frac{3}{5}$, then $\cos\theta=\frac{3}{5}$, $\sin\theta=\frac{4}{5}$. Consider the point set M comprised of $P_n(\cos 2n\theta, \sin 2n\theta)$,

$n = 1, 2, \cdots$.

For any $i, j \in \mathbf{N}^*$, we have

$$|P_iP_j|^2 = (\cos 2i\theta - \cos 2j\theta)^2 + (\sin 2i\theta - \sin 2j\theta)^2$$
$$= 2 - 2\cos 2(i-j)\theta$$
$$= 4\sin^2(i-j)\theta.$$

So, $|P_iP_j| = 2|\sin(i-j)\theta|$.

We note that $\cos\theta$, $\sin\theta \in \mathbf{Q}$, $\sin(n+1)\theta = \sin n\theta\cos\theta + \cos n\theta\sin\theta$ and $\cos(n+1)\theta = \cos n\theta\cos\theta - \sin n\theta\sin\theta$. Combining mathematical induction, we find it easy to prove that for any $n \in \mathbf{N}^*$, it is always true that $\sin n\theta$, $\cos n\theta \in \mathbf{Q}$. Therefore, the distance between any two points in M are all rational numbers.

Now we still need to prove: M is a point set with infinitely many points.

If this is not true, let us suppose M to be a finite set, then there exist $m, n \in \mathbf{N}^*$, $m \neq n$, such that $2m\theta = 2n\theta + 2k\pi$, $k \in \mathbf{Z}$. This suggests that $\theta = \alpha\pi$, $\alpha \in \mathbf{Q}$. Since $\cos\theta = \frac{3}{5} \in \mathbf{Q}$, from the conclusion above, we know that $\cos\alpha\pi \in \left\{0, \pm\frac{1}{2}, \pm 1\right\}$. However, $\cos\theta = \frac{3}{5} \notin \left\{0, \pm\frac{1}{2}, \pm 1\right\}$. This is a contradiction. So, M is a point set with infinitely many points.

For all above, there exist an infinite number of points that meet all the conditions.

33. From the conditions, we may suppose

$$P(x) = a_n(x+\beta_1)(x+\beta_2)\cdots(x+\beta_n).$$

Here $\beta_i \geqslant 1$, $i = 1, 2, \cdots, n$, and $a_n \neq 0$.

By making use of $a_0^2 + a_1a_n = a_n^2 + a_0a_{n-1}$, we can know that

$$a_n^2\left(\prod_{i=1}^n \beta_i\right)^2 + a_n^2\left(\prod_{i=1}^n \beta_i\right)\sum_{i=1}^n \frac{1}{\beta_i} = a_n^2 + \left(\prod_{i=1}^n \beta_i\right)\left(\sum_{i=1}^n \beta_i\right)a_n^2.$$

Then

$$\prod_{i=1}^{n}\beta_i-\frac{1}{\prod_{i=1}^{n}\beta_i}=\sum_{i=1}^{n}\beta_i-\sum_{i=1}^{n}\frac{1}{\beta_i}. \qquad ①$$

Next, we prove the following statement using mathematical induction, inducting with respect to n: when $\beta_i \geqslant 1$, $i=1, 2, \cdots, n$, it is always true that

$$\prod_{i=1}^{n}\beta_i-\frac{1}{\prod_{i=1}^{n}\beta_i}\geqslant\sum_{i=1}^{n}\beta_i-\sum_{i=1}^{n}\frac{1}{\beta_i}.$$

The "=" sign holds when and only when there are $n-1$ numbers equal to 1 within $\beta_1, \cdots, \beta_n$.

When $n=2$, if $\beta_1, \beta_2 \geqslant 1$, then the following relationship of equivalence holds:

$$\begin{aligned}&\beta_1\beta_2-\frac{1}{\beta_1\beta_2}\geqslant(\beta_1+\beta_2)-\left(\frac{1}{\beta_1}+\frac{1}{\beta_2}\right)\\&\Leftrightarrow(\beta_1\beta_2)^2-1\geqslant(\beta_1+\beta_2)(\beta_1\beta_2-1)\\&\Leftrightarrow(\beta_1\beta_2-1)(\beta_1-1)(\beta_2-1)\geqslant 0.\end{aligned}$$

So, when $n=2$, the above proposition holds.

Suppose the proposition holds when $n=k$, then when $n=k+1$, let $\alpha=\beta_k\beta_{k+1}$. By induction hypothesis, we may know that

$$\prod_{i=1}^{k+1}\beta_i-\frac{1}{\prod_{i=1}^{k+1}\beta_i}\geqslant\left(\sum_{i=1}^{k-1}\beta_i-\sum_{i=1}^{k-1}\frac{1}{\beta_i}\right)+\alpha-\frac{1}{\alpha},$$

where the "=" sign holds when and only when there are $k-1$ numbers equal to 1 within $\beta_1, \beta_2, \cdots, \beta_{k-1}, \alpha$.

From the $n=2$ case, we may know that

$$\alpha-\frac{1}{\alpha}=\beta_k\beta_{k+1}-\frac{1}{\beta_k\beta_{k+1}}\geqslant\beta_k+\beta_{k+1}-\frac{1}{\beta_k}-\frac{1}{\beta_{k+1}}.$$

Then,

$$\prod_{i=1}^{k+1}\beta_i - \frac{1}{\prod_{i=1}^{k+1}\beta_i} \geqslant \sum_{i=1}^{k+1}\beta_i - \sum_{i=1}^{k+1}\frac{1}{\beta_i},$$

where the "=" sign holds when and only when there are $k-1$ numbers equal to 1 within β_1, $\cdots$, β_{k-1}, α, and one of β_k and β_{k+1} is a 1. This is equivalent to "there are k numbers equal to 1 within β_1, $\cdots$, β_{k+1}."

From the above conclusion and ①, we know that polynomials in the form of $P(x) = a_n(x+1)^{n-1}(x+\beta)$, $a_n \neq 0$, $\beta \geqslant 1$ are all the polynomials that meet the conditions.

34. Let $x_1 = 1$, $x_{n+1} = P(x_n)$, $n = 1, 2, \cdots$. For a fixed $n \in \mathbf{N}^*$, $n \geqslant 2$, denote $x_n - 1 = M$. Then $x_1 \equiv 1 \equiv x_n \pmod{M}$, therefore $P(x_1) \equiv P(x_n) \pmod{M}$, i.e., $x_2 \equiv x_{n+1} \pmod{M}$. Then by making use of mathematical induction, we may prove that for any $k \in \mathbf{N}^*$, it is always true that

$$x_k \equiv x_{n+k-1} \pmod{M}. \qquad ①$$

From the condition, we know that one term from x_1, x_2, $\cdots$ is a multiple of M, therefore there exists a $r \in \mathbf{N}^*$, such that $x_r \equiv 0 \pmod{M}$. From ① we know that the sequence $\{x_k\}$ is a sequence with period $n-1$ under mod M. So, we may assume $1 \leqslant r \leqslant n-1$.

Now from $P(n) > n$ we can know that $x_1 < x_2 < \cdots < x_n$, so $x_{n-1} \leqslant x_n - 1 = M$. Furthermore, $x_r \leqslant M$. But $M \mid x_r$, so $x_r = M = x_n - 1$, and this requires that $r = n-1$, i.e., $x_n - 1 = x_{n-1}$. So $P(x_{n-1}) = x_n = x_{n-1} + 1$. Since this equation is true for any $n \geqslant 2$, as $\{x_n\}$ is a monotonically increasing sequence, we know that $P(x) = x + 1$ holds true for infinitely many distinctive positive integers.

So $P(x) = x + 1$.

35. From the conditions, we get $P(-x)^2 - 1 = P((-x)^2 - 1) = P(x^2 - 1) = P(x)^2 - 1$, so $P(x)^2 = P(-x)^2$. Now we suppose $P(x) = a_{2k+1}x^{2k+1} + a_{2k}x^{2k} + \cdots + a_1 x + a_0 (a_{2k+1} \neq 0)$. Compare the coefficients of all terms expanded for $P(x)^2$ and $P(-x)^2$ and we can get $a_{2k} = a_{2k-2} = \cdots = a_0 = 0$. So, $P(x)$ has only non-zero terms with odd-number degrees, i.e., $P(-x) = -P(x)$. Therefore, $P(0) = 0$, and,

$P(-1) = P(0^2 - 1) = P(0)^2 - 1 = -1$, $P(1) = -P(-1) = 1$.

Consider the sequence $b_1 = 1$, $b_{n+1} = \sqrt{b_n + 1}$, $n = 1, 2, \cdots$. Notice that $b_1 < b_2 = \sqrt{2}$. Now we suppose $b_n < b_{n+1}$, then $b_n + 1 < b_{n+1} + 1$, $\sqrt{b_n + 1} < \sqrt{b_{n+1} + 1}$, i. e., $b_{n+1} < b_{n+2}$. From this and by principles of mathematical induction we know that $\{b_n\}$ is an increasing sequence.

Moreover, $P(b_1) = P(1) = 1 = b_1$. Now we suppose $P(b_n) = b_n$, then

$$P(b_{n+1})^2 = P(b_{n+1}^2 - 1) + 1 = P(b_n) + 1 = b_n + 1 = b_{n+1}^2.$$

So $P(b_{n+1}) = \pm b_{n+1}$. But if $P(b_{n+1}) = -b_{n+1}$, then $P(b_{n+2})^2 = P(b_{n+1}) + 1 = 1 - b_{n+1} = 1 - \sqrt{b_n + 1} < 0$, and this is a contradiction. So $P(b_{n+1}) = b_{n+1}$. Therefore, by principles of mathematical induction we can know that for any $n \in \mathbf{N}^*$, it is always true that $P(b_n) = b_n$.

From all above, there are infinitely many distinctive real numbers x, such that $P(x) = x$. So for any x, it is always true that $P(x) = x$.

36. From (2), without loss of generality, we suppose k is the smallest positive integer number which makes $f^{(k)}(0) = 0$. If $k \geqslant 3$, then

$$\begin{aligned} |f(0)| &= |f(0) - 0| \geqslant |f^{(2)}(0) - f(0)| \geqslant \cdots \\ &\geqslant |f^{(k)}(0) - f^{(k-1)}(0)| = |f^{(k-1)}(0)|, \end{aligned}$$

while $|f^{(k-1)}(0)| = |f^{(k-1)}(0) - 0| \geqslant |f^{(k)}(0) - f(0)| = |f(0)|$.

So $|f(0)| = |f^{(k-1)}(0)|$.

If $f(0) = f^{(k-1)}(0)$, then $f(f(0)) = f^{(k)}(0) = 0$. This is a contradiction.

If $f(0) = -f^{(k-1)}(0)$, then from (1) we can know that

$$\begin{aligned} |f(0)| &= |f(0) + 0| = |f^{(k)}(0) - f^{(k-1)}(0)| \\ &\leqslant |f^{(k-1)}(0) - f^{(k-2)}(0)| \leqslant \cdots \\ &\leqslant |f^{(2)}(0) - f(0)| \leqslant |f(0) - 0| = |f(0)|. \end{aligned}$$

So, the above "inequality" signs are all replaced by "equal" signs.

Noting that all following numbers $f(0)$, $\cdots$, $f^{(k-1)}(0)$ are non-zeros. So, from the system of equations below,

$$\begin{cases} | f^{(2)}(0) - f(0) | = | f(0) |, \\ | f^{(3)}(0) - f^{(2)}(0) | = | f(0) |, \\ \cdots \\ | f^{(k-1)}(0) - f^{(k-2)}(0) | = | f(0) |. \end{cases}$$

We know that for $2 \leqslant j \leqslant k-1$, we always have $f^{(j)}(0) - f^{(j-1)}(0) = \pm f(0)$. Then $f^{(2)}(0) = 2f(0)$, $f^{(3)}(0) \in \{f(0), 3f(0)\}$, $f^{(4)}(0) \in \{2f(0), 4f(0)\}$. Conducting this recursively, we can know that $f^{(k-1)}(0)$ is a positive integral multiple of $f(0)$, which is contradictory to $f^{(k-1)}(0) = -f(0)$.

From all above, the proposition holds true.

37. We build a combinatorics model: Use $f(n)$ to represent the number of strips of $1 \times n$ we have which are made of red cubes of 1×1, blue cubes of 1×1, and white cubes of 1×2.

By direct calculations, we can know that $f(n) = \sum \frac{(i+j+k)!}{i!j!k!}$. Here the summation process is for all non-negative integer arrays (i, j, k) satisfying $i + j + 2k = n$.

On the other hand, we calculate the value of $f(n)$ by a recursive method and we get $f(1) = 2$, $f(2) = 5$. While for the strips of $1 \times (n+2)$, with lengths of $n+2$, if the first cube is red or blue, after being removed, there are altogether $f(n+1)$ strips that meet the conditions. If the first cube is white (whose length is 2), after being removed, there are altogether $f(n)$ strips that meet the conditions. So $f(n+2) = 2f(n+1) + f(n)$.

Comparing the initial values of the sequences $\{f(n)\}$ and $\{p_n\}$, noting the recursive formulas, we may know that for any $n \in \mathbf{N}^*$, it is always true that $f(n) = p_n$.

So, the proposition holds true.

38. Lemma: For any $n \in \mathbf{N}^*$, it is always true that

$$\left\{\frac{\beta_1}{2} + \frac{\beta_2}{2^2} + \cdots + \frac{\beta_n}{2^n} \middle| \beta_i \in \{-1, 1\}, i = 1, 2, \cdots, n\right\}$$

$$= \left\{\frac{j}{2^n} \middle| j \text{ is an odd number, and } |j| < 2^n\right\}.$$

The proof of the lemma can be conducted through inducting with respect to n.

When $n = 1$, the lemma obviously is true. Now we suppose the lemma holds for all positive integers less than n. Consider the case with n.

For $\beta_i \in \{-1, 1\}$, denote $j = 2^{n-1}\beta_1 + 2^{n-2}\beta_2 + \cdots + 2^0\beta_n$, then $\sum_{i=1}^{n} \frac{\beta_i}{2^i} = \frac{j}{2^n}$, and j is an odd number. Furthermore, we have

$$\left|\frac{j}{2^n}\right| = \left|\sum_{i=1}^{n} \frac{\beta_i}{2^i}\right| \leqslant \sum_{i=1}^{n} \frac{|\beta_i|}{2^i} = \sum_{i=1}^{n} \frac{1}{2^i} = 1 - \frac{1}{2^n} < 1,$$

so $|j| < 2^n$.

Conversely, for odd numbers j, and $|j| < 2^n$, we note that one of $\frac{j-1}{2}$ and $\frac{j+1}{2}$ is odd, and the other is even. We suppose j_0 is the odd number of these two numbers $\frac{j-1}{2}$ and $\frac{j+1}{2}$, then $|j_0| \leqslant \frac{1}{2}(1 + |j|) \leqslant 2^{n-1} + \frac{1}{2}$. Concerning also that j_0 is an odd number, we know $|j_0| < 2^{n-1}$. Hence from induction hypothesis we know that there exist β_1, β_2, $\cdots$, $\beta_{n-1} \in \{-1, 1\}$, such that $\frac{\beta_1}{2} + \frac{\beta_2}{2^2} + \cdots + \frac{\beta_{n-1}}{2^{n-1}} = \frac{j_0}{2^{n-1}}$.

Let $\beta_n = j - 2j_0$, then $\beta_n \in \{-1, 1\}$, and that

$$\sum_{i=1}^{n} \frac{\beta_i}{2^i} = \frac{j_0}{2^{n-1}} + \frac{j - 2j_0}{2^n} = \frac{j}{2^n}.$$

So, the lemma is proved.

(1) From the conclusion of the lemma and the structure of all elements in A_n, we may know that $A_n = \left\{1 + \frac{j}{2^{\lfloor \frac{n}{2} \rfloor}} + \frac{k}{2^{\lceil \frac{n}{2} \rceil}}\sqrt{2} \;\middle|\; j, k \text{ is an odd number, and } |j| < 2^{\lfloor \frac{n}{2} \rfloor},\ k < 2^{\lceil \frac{n}{2} \rceil}\right\}$. So, from $\sqrt{2}$ being an irrational number we may know that $|A| = 2^{\lfloor \frac{n}{2} \rfloor} \cdot 2^{\lceil \frac{n}{2} \rceil} = 2^n$.

(2) Denote $S = \sum_{\substack{a, b \in A_n \\ a < b}} ab$, then $S = \frac{1}{2}\left(\left(\sum_{a \in A_n} a\right)^2 - \sum_{a \in A_n} a^2\right)$.

We pair the elements, $1+\frac{j}{2^{\lfloor\frac{n}{2}\rfloor}}+\frac{k}{2^{\lceil\frac{n}{2}\rceil}}\sqrt{2}$ and $1-\frac{j}{2^{\lfloor\frac{n}{2}\rfloor}}-\frac{k}{2^{\lceil\frac{n}{2}\rceil}}\sqrt{2}$, in A_n and sum them up. Making use of the conclusion of (1), we can know that $\sum_{a\in A_n} a = 2^n$.

Furthermore, we make use of the conclusion: if X, Y are both finite sets, and $\sum_{x\in X} x = \sum_{y\in Y} y = 0$, then

$$\sum_{x\in X}\sum_{y\in Y}(1+x+y)^2 = |X|\cdot|Y| + |Y|\cdot\sum_{x\in X}x^2 + |X|\cdot\sum_{y\in Y}y^2.$$

Concerning the structure of A_n, we can have

$$\begin{aligned}\sum_{a\in A_n}a^2 &= \sum_{\substack{j\text{ is odd}\\ |j|<2^{\lfloor\frac{n}{2}\rfloor}}}\sum_{\substack{k\text{ is odd}\\ |k|<2^{\lceil\frac{n}{2}\rceil}}}\left(1+\frac{j}{2^{\lfloor\frac{n}{2}\rfloor}}+\frac{k}{2^{\lceil\frac{n}{2}\rceil}}\sqrt{2}\right)^2\\ &= 2^{\lceil\frac{n}{2}\rceil}\cdot 2^{\lfloor\frac{n}{2}\rfloor}+\sum_{\substack{j\text{ is odd}\\ |j|<2^{\lfloor\frac{n}{2}\rfloor}}}\frac{j^2\cdot 2^{\lceil\frac{n}{2}\rceil}}{2^{2\lfloor\frac{n}{2}\rfloor}}+\sum_{\substack{k\text{ is odd}\\ |k|<2^{\lceil\frac{n}{2}\rceil}}}\frac{2k^2\cdot 2^{\lfloor\frac{n}{2}\rfloor}}{2^{2\lceil\frac{n}{2}\rceil}}\\ &= 2^n+\frac{1}{3}\left(\frac{2^{\lceil\frac{n}{2}\rceil}\cdot 2^{\lfloor\frac{n}{2}\rfloor}(2^{2\lfloor\frac{n}{2}\rfloor}-1)}{2^{2\lfloor\frac{n}{2}\rfloor}}+\frac{2^{\lfloor\frac{n}{2}\rfloor}\cdot 2^{\lceil\frac{n}{2}\rceil}(2^{2\lceil\frac{n}{2}\rceil}-1)}{2^{2\lceil\frac{n}{2}\rceil-1}}\right)\\ &= 2^n+\frac{2^n}{3}\left(3-\frac{1}{2^{2\lfloor\frac{n}{2}\rfloor}}-\frac{1}{2^{2\lceil\frac{n}{2}\rceil-1}}\right)\\ &= 2^{n+1}-1.\end{aligned}$$

From all above, the answer to (1) is 2^n, and the answer to (2) is $\frac{1}{2}(2^{2n}-2^{n+1}+1)$.

39. Lemma: Suppose $n\in\mathbf{N}^*$, and we may express it in the form of a combination of several 3s and 4s. The ordered divisions of n can be arranged in form of a matrix, then x_n is the sum of all entries on the first column of the matrix.

For example, when $n=15$, we may get the matrix as follows.

4, 4, 4, 3
4, 4, 3, 4
4, 3, 4, 4
3, 4, 4, 4
3, 3, 3, 3, 3

Directly using the recursive formula, we can calculate and get $x_{15} = 18$, the sum of all entries on the first column of the above matrix.

The proof of the lemma: We will prove through inducting with respect to n. When $n = 1, 2, 3, 4$, we may know that the proposition holds by direct verification. Now we suppose the lemma is true for all subscripts less than $n(\geqslant 5)$. We consider the case with n.

According to the ordered divisions of n into combinations of 3 and 4, we may classify the divisions into two categories by the last term being 3 or 4. For the ordered divisions with last term 3, we remove the last term 3 and we get all the ordered divisions of $n - 3$. For the ordered divisions with last term 4, we remove the last term 4 and we get all the ordered divisions of $n - 4$. Concerning $n \geqslant 5$, if the division is still possible, then at least there are two terms. Then we know that the matrix comprised of the ordered divisions of n has the sum of all entries on the first column $x_{n-3} + x_{n-4}$ (here the induction hypothesis is used). So, the lemma is true for n, and the proposition is proved.

Going back to the original question, we know when $p = 2, 3$, $x_p = 0$, the proposition holds true. For prime numbers $p(\geqslant 5)$, suppose the matrix we get by expressing p into ordered divisions of 3 and 4 is M. Then the length l for each row of M satisfies $\frac{p}{4} \leqslant l \leqslant \frac{p}{3}$.

We analyze the sub-matrix $\boldsymbol{T}$ made of all the rows of the same length in M: Suppose the sum of all elements on the first column of $\boldsymbol{T}$ is S. Since for every row of $\boldsymbol{T}$, 3 and 4 must appear at the same time (when only 3 or 4 appears, then p is a multiple of 3 or 4, and then is not prime). Hence the division of p by exchanging positions of 3 and 4 in this particular row is another row of $\boldsymbol{T}$. This suggests: the sums of elements on any two columns of $\boldsymbol{T}$ should be the same (since if the numbers in corresponding positions of these two columns are different, then there must be numbers obtained by one row intersecting with these two columns that are just a swap of the numbers mentioned before). Denote the sum of each column of $\boldsymbol{T}$ is S. Suppose the number of columns of $\boldsymbol{T}$ is l, then the sum of all numbers in $\boldsymbol{T}$ is sl, while the sums of all numbers in each row of $\boldsymbol{T}$ are all p, so sl is a multiple of p. As $\frac{p}{4} \leqslant l \leqslant \frac{p}{3}$, we know that $p \mid s$.

By above discussions, we can know that for the sum of all numbers on the first column of M, it is also a multiple of p, i. e., $p \mid x_p$. The proposition holds.

40. Suppose a_1, $\cdots$, a_n are the positive integers that meet the requirements. Then by (1) we know that for $1 \leqslant k \leqslant n$, it's always true that $a_k > a_{k-1}$ (otherwise the left-hand side $\geqslant 1$, the right-hand side < 1). Therefore, it is true that $a_k > 1$. Hence, from (2) we know that $\frac{a_{k-1}}{a_k(a_k - 1)} \geqslant \frac{a_k}{a_{k+1} - 1}$, i.e., $\frac{a_{k-1}}{a_k - 1} - \frac{a_{k-1}}{a_k} \geqslant \frac{a_k}{a_{k+1} - 1}$, so $\frac{a_{k-1}}{a_k} \leqslant \frac{a_{k-1}}{a_k - 1} - \frac{a_k}{a_{k+1} - 1}$. Summing up for $k = i + 1$, $\cdots$, $n - 1$, we have

$$\frac{a_i}{a_{i+1}} + \cdots + \frac{a_{n-1}}{a_n} \leqslant \frac{a_i}{a_{i+1} - 1} - \frac{a_{n-1}}{a_n - 1} + \frac{a_{n-1}}{a_n} < \frac{a_i}{a_{i+1} - 1}. \quad ①$$

Let $i = 0$ in ①, and by (1) we have $\frac{1}{a_1} \leqslant \frac{99}{100} = \sum_{i=0}^{n-1} \frac{a_i}{a_{i+1}} < \frac{1}{a_1 - 1}$, so $a_1 = 2$. Similarly, take $i = 1$ in ①, also concerning (1), we have

$$\frac{1}{a_2} \leqslant \frac{1}{a_1}\left(\frac{99}{100} - \frac{1}{a_1}\right) < \frac{1}{a_2 - 1},$$

and we have $\frac{1}{a_2} \leqslant \frac{49}{200} < \frac{1}{a_2 - 1}$, then we know $a_2 = 5$. Repeating these discussions, and we take $i = 2$, 3 in ①, then we can get $a_3 = 56$, $a_4 = 25 \times 56^2 = 78\,400$. So,

$$\frac{1}{a_5} \leqslant \frac{1}{a_4}\left(\frac{99}{100} - \frac{1}{2} - \frac{2}{5} - \frac{5}{56} - \frac{56}{25 \times 56^2}\right) = 0,$$

and a_5 does not exist.

By all of above, only when $n = 4$ does this kind of sequence exist, and the corresponding a_1, a_2, a_3, a_4 are 2, 5, 56, 78 400.

41. Let $x_n = ny_n$, $n = 2, 3, \cdots$, then $x_2 = 2$, $x_3 = 3$, and for $n \geqslant 3$, we have $(n-2)x_{n+1} = (n^2 - n - 1)x_n - (n-1)^2 x_{n-1}$, i.e.,

$$\frac{x_{n+1}-x_n}{n-1}=(n-1)\cdot\frac{x_n-x_{n-1}}{n-2}. \quad ①$$

Let us use ① and conduct reasoning recursively, then we can get:

$$\frac{x_{n+1}-x_n}{n-1}=(n-1)\cdot\frac{x_n-x_{n-1}}{n-2}=(n-1)(n-2)\cdot\frac{x_{n-1}-x_{n-2}}{n-3}$$
$$=\cdots=(n-1)\cdots 2\cdot\frac{x_3-x_2}{1}=(n-1)!,$$

and we get $x_{n+1}-x_n=n!-(n-1)!$. By using telescoping series, we know that

$$x_n=x_2+\sum_{k=2}^{n-1}(x_{k+1}-x_k)=x_2+\sum_{k=2}^{n-1}(k!-(k-1)!)$$
$$=x_2+(n-1)!-1=(n-1)!+1.$$

From Wilson's Theorem which claims $n\mid(n-1)!+1$ if and only if n is a prime number, we know that the necessary and sufficient condition for $y_n\in\mathbf{Z}$ is that n is a prime number.

42. Lemma: Suppose $n, k\in\mathbf{N}^*$, $n\geqslant kp$, then

$$a_n=\sum_{i=0}^{k}C_k^i a_{n-i(p-1)-k}. \quad ①$$

Induct with respect to k. When $k=1$, ① is actually $a_n=a_{n-1}+a_{n-p}$, so ① is true for $k=1$.

Now we suppose ① is true for k, then we consider the case with $k+1$. So, $n\geqslant(k+1)p$, and the minimum value of the subscripts $n-i(p-1)-k(0\leqslant i\leqslant k)$ is reached when $i=k$, and the minimum value is $n-kp\geqslant p$. So, every term of the following summation process can apply the recursive formula in the condition:

By induction hypothesis, when $n\geqslant(k+1)p$, we have

$$a_n=\sum_{i=0}^{k}C_k^i a_{n-i(p-1)-k}$$
$$=\sum_{i=0}^{k}C_k^i(a_{n-i(p-1)-k-1}+a_{n-i(p-1)-k-p})$$
$$=C_k^0a_{n-k-1}+\sum_{i=1}^{k}C_k^ia_{n-i(p-1)-k-1}+\sum_{i=0}^{k-1}C_k^ia_{n-i(p-1)-k-p}+C_k^ka_{n-(k+1)p}$$

$$= C_{k+1}^0 a_{n-(k+1)} + \sum_{i=0}^{k-1} C_k^{i+1} a_{n-(i+1)(p-1)-(k+1)}$$

$$+ \sum_{i=0}^{k-1} C_k^i a_{n-(i+1)(p-1)-(k+1)} + C_{k+1}^{k+1} a_{n-(k+1)p}$$

$$= C_{k+1}^0 a_{n-(k+1)} + \sum_{i=0}^{k-1} (C_k^{i+1} + C_k^i) a_{n-(i+1)(p-1)-(k+1)} + C_{k+1}^{k+1} a_{n-(k+1)p}$$

$$= \sum_{i=0}^{k+1} C_{k+1}^i a_{n-(i+1)(p-1)-(k+1)}.$$

For the last step, $C_k^{i+1} + C_k^i = C_{k+1}^{i+1}$ is used. So, ① holds true for $k+1$, and the lemma is proved.

Next, we are going to deal with the original problem by making use of the lemma. When $n \geqslant p^2$, let $k = p$ in the lemma, then we have

$$a_n = \sum_{i=0}^{p} C_p^i a_{n-i(p-1)-p}.$$

It is well-known that when $1 \leqslant i \leqslant p-1$, we have $C_p^i \equiv 0 \pmod p$. So, $a_n \equiv a_{n-p} + a_{n-p^2} \pmod p$, concerning $a_n = a_{n-1} + a_{n-p}$, we have $a_{n-1} \equiv a_{n-p^2} \pmod p$. This suggests that for any $t \geqslant p^2 - 1$, we have

$$a_t \equiv a_{t+p^2-1} \pmod p.$$

Since $p^3 = p(p^2-1) + p$, $a_{p^3} = a_{p+p(p^2-1)} \equiv a_p \pmod p$, and also since $a_p = a_0 + a_{p-1} = p-1$, then we have $a_{p^3} \equiv p-1 \pmod p$, i.e., the remainder is $p-1$ when a_{p^3} is divided by p.

43. For the sake of convenience, denote $m = n-1$, $b_i = a_i + 1$, then $1 \leqslant a_0 \leqslant 2m$, and

$$a_{i+1} = \begin{cases} 2a_i, & \text{if } a_i \leqslant m, \\ 2a_i - (2m+1), & \text{if } a_i > m. \end{cases}$$

This suggests that $a_{i+1} \equiv 2a_i \pmod{2m+1}$, and $1 \leqslant a_i \leqslant 2m$, $i = 1, 2, \cdots$.

(1) The $p(2, 2^k)$ and $p(2, 2^k+1)$ in question are equivalent to finding $p(1, 2^k-1)$ and $p(1, 2^k)$ for $\{a_i\}$. The former one is equivalent to finding the smallest $l \in \mathbf{N}^*$, such that $2^l \equiv 1 \pmod{2(2^k-1)+1}$, and the latter one is equivalent to finding the smallest $l \in \mathbf{N}^*$,

such that $2^l \equiv 1 \pmod{2^{k+1}+1}$.

Obviously, $2^{k+1} \equiv 1 \pmod{2(2^k-1)+1}$. For $1 \leqslant t \leqslant k$, it is always true that

$$1 \leqslant 2^t - 1 < 2^{k+1} - 1 = 2(2^k - 1) + 1,$$

so $p(1, 2^k - 1) = k + 1$.

Since $2^{2(k+1)} \equiv 1 \pmod{2^{k+1}+1}$, then $p(1, 2^k) \mid 2(k+1)$. While for $1 \leqslant t \leqslant k+1$, it is always true that $1 \leqslant 2^t - 1 < 2^{k+1} + 1$, then $p(1, 2^k) > k+1$, so $p(1, 2^k) = 2(k+1)$.

So, for $\{b_i\}$, we have $p(2, 2^k) = k+1$, and that $p(2, 2^k + 1) = 2(k+1)$.

(2) We still go to $\{a_i\}$ to proceed with the discussion. It is required to prove: $p(a_0, m) \mid p(1, m)$. Now we suppose $p(1, m) = t$, then $2^t \equiv 1 \pmod{2m+1}$, therefore, $2^t a_0 \equiv a_0 \pmod{2m+1}$. This suggests that $p(a_0, m) \mid t$ (here we make use of some property of "order" in basic number theory). That is to say, $p(a_0, m) \mid p(1, m)$, and the proposition is true.

44. First, we set up a lemma: for any $\alpha \in (0, 1)$, there exist two points on a broken line such that they share the same ordinates, and their abscissae differ by α or $1-\alpha$.

As a matter of fact, suppose Γ is the given broken line in question, Γ_1 is the broken line obtained by shifting Γ to the left by α units, and Γ_2 is the broken line obtained by shifting Γ to the right by $1-\alpha$ units. It is easy to get that there is at least one point of intersection for Γ and $\Gamma_1 \cup \Gamma_2$, and this is the result that the lemma needs (as shown by Figure 7, we will know that there will be points of intersection for Γ and $\Gamma_1 \cup \Gamma_2$, if we start from the highest and lowest points of Γ).

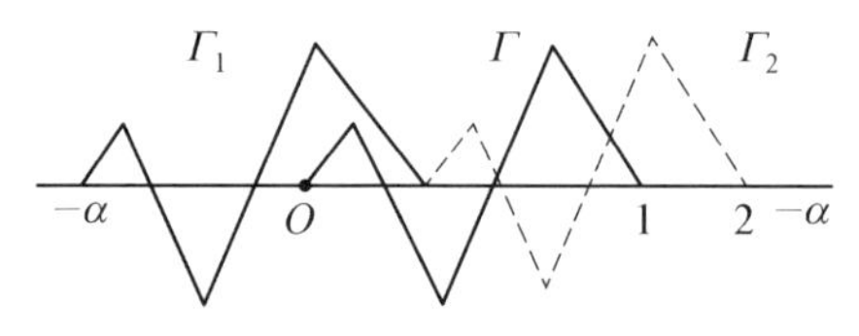

Figure 7

Next, we make use of the lemma to prove the conclusion we need.

Take $\alpha = \frac{1}{2}$, and we may know that when $n = 2$, the conclusion holds; Take $\alpha = \frac{1}{3}$, and we have two points A, B on the broken line, such that $AB \mathbin{/\!/}$ the x-axis, and $|AB| = \frac{1}{3}$, or $|AB| = \frac{2}{3}$. If $|AB| = \frac{1}{3}$, then $n = 3$ already holds. If $|AB| = \frac{2}{3}$, then we consider the sub-broken line joining A with B. Referring to the lemma and the conclusion with $n = 2$, we may know that there exist points C, D on the sub-broken line, such that $CD \mathbin{/\!/} AB$, and $|CD| = \frac{1}{2}|AB| = \frac{1}{3}$. Hence, the conclusion holds true when $n = 3$. Conduct the reasoning likewise, together with mathematical induction, we may know that the conclusion holds true for any $n \geqslant 2$.

45. We first use mathematical induction to prove the existence.

When $n = 1$, obviously the method exists. Suppose when n, there is a method T that meets the conditions. We now consider the case with $n+1$. Then we first put the n balls into the white boxes labeled $1, 2, \cdots, n$, and we suppose after this placement, the smallest labeling number of the empty box is $i(1 \leqslant i \leqslant n)$, then we follow the method described below to put in the $n+1$ th ball: We take one ball respectively from the white boxes labeled $1, 2, \cdots, i-1$ and put them in the box labeled as i, and we put the $n+1$ th ball also in the box labeled as i. It is easy to know that this way of placement meets the conditions.

Then we use mathematical induction to prove that the method of placement is unique, inducting still with respect to n.

When $n = 1, 2$, the uniqueness obviously holds. Suppose for $n(\geqslant 2)$, there is only one method of placement, denoted by T.

It is easy to know when $n + 1 \geqslant 3$, among all the methods of placement, the $n+1$ th white box must be empty. Then, when $n+1$, there exist two methods of placement T_1 and T_2. We notice that the

$n+1$ th white box, which is empty, can be removed, and after one-step operation for T_1 and T_2, there are n balls in the white box, and the number of white boxes is n, so they both become T.

Suppose the labeled number of white boxes that were operated under T_1 and T_2 respectively are i_1, i_2. If $i_1 > i_2$, then after the first operation under T_1, the white box labeled i_2 has at least one ball, while after the first operation under T_2, there is no ball in the white box labeled i_2. They cannot both change into T, so $i_1 \leqslant i_2$. By the same reasoning, $i_2 \leqslant i_1$, i. e., $i_1 = i_2$. Thus, under T_1 and T_2, the numbers of balls in the white boxes whose labels are bigger than i_1 are the same. The numbers of balls in the white boxes whose labels are less than i_1 are also the same (otherwise, T_1 and T_2, after one operation, they cannot both change into T.) So, the numbers of balls in the boxes labeled i_1 should also be the same, therefore $T_1 = T_2$.

This suggests that there exists a unique method of placement that meets the conditions.

46. (1) We use 0, 1, 2 to express A, B and C, respectively. Under modulus 3, we watch the changing status of the sequence R_0, R_1, $\cdots$.

Let $R_j = (x_1, \cdots, x_n)$, $R_{j+1} = (y_1, y_2, \cdots, y_n)$, then for $1 \leqslant i \leqslant n$, it is always true that $y_i \equiv -(x_i + x_{i+1}) \pmod 3$.

If n is an even number, take $R_0 = (1, 2, 1, 2, \cdots, 1, 2)$. Then for any $m \geqslant 1$, it is always true that $R_m = (0, 0, \cdots, 0, 0)$. So at this time, there does not exist any positive integer m that meets the requirements. If n is an odd number, since there are at most 3^n distinct n-element number arrays in form of $(x_1, \cdots, x_n)$, then for any R_0, there exists an $m_{R_0} \in \mathbf{N}^*$, and $k \in \mathbf{N}$, such that $R_k = R_{m_{R_0}+k}$. We will prove that if $k \geqslant 1$, then $R_{k-1} = R_{m_{R_0}+k-1}$ (therefore by inferring likewise we get $R_0 = R_{m_{R_0}}$).

As a matter of fact, suppose $R_{k-1} = (x_1, \cdots, x_n)$, $R_{m_{R_0}+k-1} = (y_1, \cdots, y_n)$, then from $R_k = R_{m_{R_0}+k}$, we can know that $-(x_i + x_{i+1}) \equiv -(y_i + y_{i+1}) \pmod 3$. So

$$\sum_{j=1}^{n}(-1)^j(x_j+x_{j+1}) \equiv \sum_{j=1}^{n}(-1)^j(y_j+y_{j+1}) \pmod 3.$$

Concerning also n is an odd number, we know that $-x_1+(-1)^n x_{n+1} \equiv -y_1+(-1)^n y_{n+1} \pmod 3$, i.e., $-2x_1 \equiv -2y_1 \pmod 3$, $x_1 \equiv y_1 \pmod 3$. So, $x_1 = y_1$. By the same reasoning, we can prove that for $i \in \{2, \cdots, n\}$, it is always true that $x_i = y_i$, so $R_{k-1} = R_{m_{R_0}+k-1}$.

By above we know that for any R_0, there exists an m_{R_0} such that $R_0 = R_{m_{R_0}}$. Then, while R_0 changes, we take the least common multiple m of all m_{R_0}. Then for any R_0, it is always true $R_0 = R_m$.

From all above, there exists an m that meets all conditions if and only if n is an odd number.

(2) For $n = 3^k$, $k \in \mathbf{N}^*$, the smallest value of m that meets the conditions (set in (1)) should be $m = 3^k$.

As a matter of fact, for any $R_0 = (x_1, \cdots, x_n)$, suppose $R_{3^k} = (y_1, \cdots, y_n)$, then from the relation we inferred before under modulus 3, it is easy to know that $y_p \equiv -\sum_{i=0}^{3^k} C_{3^k}^i x_{i+p} \pmod 3$, where the subscript of x_{i+p} takes on values under the meaning of modulus n, where $p = 1, 2, \cdots, n$. We also notice that for $1 \leqslant i \leqslant 3^k - 1$, $C_{3^k}^i \equiv 0 \pmod 3$, so $y_p \equiv -x_p - x_{3^k+p} = -2x_p \equiv x_p \pmod 3$, therefore, $R_{3^k} = R_0$.

On the other hand, suppose $R_0 = (0, 0, \cdots, 0, 1)$, then for $0 < m < 3^k$, the $3^k - m$ th component of R_m is not equal to 0. So, the minimum value of m that satisfies (1) is 3^k.

Solutions to Exercise Set 2

1. We make the proposition more general by changing 2011 into n, and we prove that the proposition holds true for the n-element set S and $0 \leqslant N \leqslant 2^n$.

When $n = 1$, the subsets of S are S and $\varnothing$ only. For $0 \leqslant N \leqslant 2$, we dye any N subsets into white, and the others into black, then we know that the proposition holds true.

We suppose the proposition holds true for n, then we consider the case with $n+1$. We partition the subsets of S into the part containing a_{n+1} and the part not containing a_{n+1}. Thus, $S = \{a_1, \cdots, a_{n+1}\}$. Suppose the subsets of S that do not contain a_{n+1} include $A_1, \cdots, A_{2^n}$, the subsets of S that contain a_{n+1} include $B_1, \cdots, B_{2^n}$, where

$$B_i = A_i \cup \{a_{n+1}\},\ 1 \leqslant i \leqslant 2^n.$$

If $0 \leqslant N \leqslant 2^n$, then by induction hypothesis, we dye N sets from $A_1, \cdots, A_{2^n}$ into white color, and others into black color. We dye all B_i into black color after meeting all conditions set by the question. Then we know that the proposition is true for $n+1$. If $2^n < N \leqslant 2^{n+1}$, then suppose $N = 2^n + k$, where $0 < k \leqslant 2^n$. We use the method we applied in induction hypothesis for $A_1, \cdots, A_{2^n}$, which is to dye k of them into white color, and others into black color such that all conditions are met. Then we dye all B_i into white color.

Generalizing, we know that the proposition is true for all n and then of course true for $n = 2011$.

2. We extend 2048 into the case 2^n. Namely, we prove: for any $n \in \mathbf{N}^*$, we place $+1$ and -1, totaling 2^n terms, on a circle. By operations described by the question, after a limited number of operations, all the numbers will become $+1$.

When $n = 1$, by the given conditions, we can get the following sequence for operations:

$$(+1, -1) \rightarrow (-1, -1) \rightarrow (+1, +1).$$

Then we know that the proposition holds true for $n = 1$.

Suppose the proposition holds true for n, then for the case $n+1$, we use $x_1, x_2, \cdots, x_{2^{n+1}}$ to express the 2^{n+1} numbers permuted on the circle. Then, we have the following sequence for operations:

$$(x_1, x_2, \cdots, x_{2^{n+1}}) \rightarrow (x_1x_2, x_2x_3, \cdots, x_{2^{n+1}}x_1) \rightarrow$$
$$(x_1x_3, x_2x_4, \cdots, x_{2^{n+1}}x_2).$$

We "combine" the two operations into one operation. Then we know that if the 2^n numbers $(x_1, x_3, \cdots, x_{2^{n+1}-1})$ and $(x_2, x_4, \cdots, x_{2^{n+1}})$ on the circle can all be changed into $+1$ after a limited number of operations, the proposition is proved. Since this requirement is exactly the induction hypothesis, the proposition is hence true.

3. Without loss of generality, we suppose $x_1, \cdots, x_n \geqslant 0$. When $n = 1$, $\frac{x_1}{1+x_1^2} \leqslant \frac{1}{2} < 1$, so the proposition is true for $n = 1$. If we suppose the proposition holds true for n, then we consider the case $n+1$. Let $y_{i-1} = \frac{x_i}{\sqrt{1+x_1^2}}$, $i = 2, 3, \cdots, n+1$, then

$$\sum_{i=1}^{n+1} \frac{x_i}{1+x_1^2+\cdots+x_i^2} = \frac{x_1}{1+x_1^2} + \frac{1}{\sqrt{1+x_1^2}} \sum_{i=1}^{n} \frac{y_i}{1+y_1^2+\cdots+y_i^2}$$
$$< \frac{x_1}{1+x_1^2} + \frac{\sqrt{n}}{\sqrt{1+x_1^2}}.$$

Now we suppose $x_1 = \tan\alpha$, $0 \leqslant \alpha < \frac{\pi}{2}$, then

$$\frac{x_1}{1+x_1^2} + \frac{\sqrt{n}}{\sqrt{1+x_1^2}} = \sin\alpha\cos\alpha + \sqrt{n}\cos\alpha \leqslant \sin\alpha + \sqrt{n}\cos\alpha$$
$$= \sqrt{n+1}\sin(\alpha+\varphi) \leqslant \sqrt{n+1},$$

Where $\varphi = \arctan\sqrt{n}$.

So, the proposition is true for $n+1$. QED.

Explanation. There is a very smart solution to this problem. Let $x_0 = 0$, then by Cauchy-Schwartz Inequality we know that

$$\left(\sum_{i=1}^{n} \frac{x_i}{1+x_1^2+\cdots+x_i^2}\right)^2$$
$$\leqslant n\sum_{i=1}^{n} \frac{x_i^2}{(1+x_1^2+\cdots+x_i^2)^2}$$
$$\leqslant n\sum_{i=1}^{n} \frac{x_i^2}{(1+x_0^2+\cdots+x_{i-1}^2)(1+x_0^2+\cdots+x_i^2)}$$

$$= n\sum_{i=1}^{n}\left(\frac{1}{1+x_0^2+\cdots+x_{i-1}^2}-\frac{1}{1+x_0^2+\cdots+x_i^2}\right)$$
$$= n\left(1-\frac{1}{1+x_0^2+\cdots+x_n^2}\right)$$
$$< n.$$

So the original inequality is true.

4. First, we prove a lemma:

$$\sum_{(\varepsilon_1,\cdots,\varepsilon_n)} |\varepsilon_1 z_1+\cdots+\varepsilon_n z_n|^2 = 2^n \cdot \sum_{k=1}^{n} |z_k|^2,$$

where the summation process is meant for all possible vectors $(\varepsilon_1, \cdots, \varepsilon_n)$, where $\varepsilon_i \in \{-1, 1\}$.

When $n=1$, it is obvious that the lemma is true. When $n=2$, we notice that

$$|z_1-z_2|^2+|z_1+z_2|^2 = (z_1-z_2)(\bar{z}_1-\bar{z}_2)+(z_1+z_2)(\bar{z}_1+\bar{z}_2)$$
$$= 2(z_1\bar{z}_1+z_2\bar{z}_2) = 2(|z_1|^2+|z_2|^2),$$

(This conclusion can be interpreted as the sum of squares for lengths of diagonals of a parallelogram is equal to the sum of squares for its four sides.) and by this, we know that the lemma is true for $n=2$.

Now we suppose the lemma is true for n, then from

$$\sum_{(\varepsilon_1,\cdots,\varepsilon_{n+1})} |\varepsilon_1 z_1+\cdots+\varepsilon_{n+1} z_{n+1}|^2$$
$$= \sum_{(\varepsilon_1,\cdots,\varepsilon_n)} (|\varepsilon_1 z_1+\cdots+\varepsilon_n z_n+z_{n+1}|^2+|\varepsilon_1 z_1+\cdots+\varepsilon_n z_n-z_{n+1}|^2)$$
$$= 2\sum_{(\varepsilon_1,\cdots,\varepsilon_n)} (|\varepsilon_1 z_1+\cdots+\varepsilon_n z_n|^2+|z_{n+1}|^2)$$
$$= 2^{n+1}|z_{n+1}|^2+2\sum_{(\varepsilon_1,\cdots,\varepsilon_n)} |\varepsilon_1 z_1+\cdots+\varepsilon_n z_n|^2$$
$$= 2^{n+1}|z_{n+1}|^2+2^{n+1}\sum_{k=1}^{n}|z_k|^2$$
$$= 2^{n+1}\sum_{k=1}^{n+1}|z_k|^2.$$

We know that the lemma is true for $n+1$. So for any $n \in \mathbf{N}^*$, the

Lemma is true.

Coming back to the original problem, by conditions we know that

$$\sum_{(\varepsilon_1, \cdots, \varepsilon_n)} |\varepsilon_1 z_1 + \cdots + \varepsilon_n z_n|^2 \leqslant \sum_{(\varepsilon_1, \cdots, \varepsilon_n)} |\varepsilon_1 \omega_1 + \cdots + \varepsilon_n \omega_n|^2.$$

Then by the lemma, we know that $2^n \sum_{k=1}^{n} |z_k|^2 \leqslant 2^n \sum_{k=1}^{n} |\omega_k|^2$, then the proposition is proved.

5. A polynomial that obviously satisfies the condition is $P = x_1 x_2 \cdots x_n$. If we are able to prove one term of $P(x_1, \cdots, x_n)$ is a multiple of $x_1 x_2 \cdots x_n$ (i.e., $x_1, \cdots, x_n$ all appear in this term), then the degree of P is no less than n.

Now we prove the strengthened conclusion: One term of $P(x_1, \cdots, x_n)$ is a multiple of $x_1 x_2 \cdots x_n$.

When $n = 1$, from the condition $P(1) > 0$, $P(-1) < 0$, then we know that $P(x_1)$ is not a constant, and one term is a multiple of x_1. The proposition holds.

Suppose the proposition is true for any polynomial with $n - 1$ variables that meets the conditions. We consider the case with n.

For $P(x_1, x_2, \cdots, x_n)$ that meets the conditions, we let

$$\begin{aligned} & Q(x_1, x_2, \cdots, x_{n-1}) \\ = & \frac{1}{2}[P(x_1, x_2, \cdots, x_{n-1}, 1) - P(x_1, \cdots, x_{n-1}, -1)], \end{aligned}$$

which is the sum of all coefficients of the odd-degree terms of x_n, when we treat P as a polynomial of x_n (other variables $x_1, \cdots, x_{n-1}$ are all treated as constants).

Since when $x_1, \cdots, x_{n-1}$ are all replaced by $+1$ or -1, if the number of -1 is an even number, then $P(x_1, \cdots, x_{n-1}, 1) > 0$, $P(x_1, \cdots, x_{n-1}, -1) < 0$, so $Q(x_1, \cdots, x_{n-1}) > 0$; similarly, if the number of -1 is an odd number, then $Q(x_1, \cdots, x_{n-1}) < 0$. By making use of induction hypothesis we may know that one term from $Q(x_1, \cdots, x_{n-1})$ is a multiple of $x_1 x_2 \cdots x_{n-1}$. Note that $P(x_1, \cdots, x_n)$ originates from summing up all the terms by first multiplying every term in $Q(x_1, \cdots, x_{n-1})$ by a certain odd-number power of x_n. So, one term of $P(x_1, \cdots, x_n)$ is a multiple of $x_1 x_2 \cdots x_n$.

From all above, the proposition holds true.

6. When $n = 1$, $m_1 = a_1$. If $\mu \geqslant a_1$, then the subscript k that satisfies $m_k > \mu$ does not exist, and at this time, the proposition is obviously true. If $\mu < a_1$, then there is exactly one subscript k that meets the requirements. From $1 < \frac{a_1}{\mu}$ we know that the proposition holds true.

Now we suppose the proposition holds true for $1, 2, \cdots, n-1 (n \geqslant 2)$. Suppose for the case n, r is the number of subscripts k that satisfies $m_k > \mu$. If $m_k \leqslant \mu$, then for the sequence $a_1, \cdots, a_{n-1}$, the number of subscipts k that satisfies $m_k > \mu$ is also r. Then from induction hypothesis, we can know that $r < \frac{a_1 + \cdots + a_{n-1}}{\mu} \leqslant \frac{a_1 + \cdots + a_n}{\mu}$. The proposition is true for n.

If $m_n > \mu$, then there exists $i \in \{1, 2, \cdots, n\}$ such that $\frac{a_{n-i+1} + \cdots + a_n}{i} > \mu$. For this i, concerning the sequence $a_1, a_2, \cdots, a_{n-i}$, there are at least $r - i$ subscripts k that satisfies $m_k > \mu$. Hence, from induction hypothesis, we know that $r - i < \frac{a_1 + \cdots + a_{n-i}}{\mu}$.

Then, $(a_1 + \cdots + a_{n-i}) + (a_{n-i+1} + \cdots + a_n) > (r-i)\mu + i\mu = r\mu$, so $r < \frac{a_1 + a_2 + \cdots + a_n}{\mu}$.

The proposition is proved.

7. Compare to the second proof for the Arithmetic Mean-Geometric Mean Inequality in Section 10. We use the method in that proof to prove the widely used Jensen's Inequality.

When $n = 1, 2$, it is obvious that the inequality is true.

Now we suppose the inequality holds for $n = 2^k (k \in \mathbf{N}^*)$. Then from the definition of f, we may know that

$$f\left(\frac{x_1+\cdots+x_{2^{k+1}}}{2^{k+1}}\right)\leqslant\frac{1}{2}\left(f\left(\frac{x_1+\cdots+x_{2^k}}{2^k}\right)+f\left(\frac{x_{2^k+1}+\cdots+x_{2^{k+1}}}{2^k}\right)\right)$$

$$\leqslant\frac{1}{2}\left(\frac{1}{2^k}\sum_{j=1}^{2^k}f(x_j)+\frac{1}{2^k}\sum_{j=1}^{2^k}f(x_{2^k+j})\right)$$

$$=\frac{1}{2^{k+1}}\sum_{j=1}^{2^{k+1}}f(x_j).$$

Hence, the inequality is true for any $n=2^k(k\in\mathbf{N}^*)$.

More generally, for $n\in\mathbf{N}^*$ ($n\geqslant 3$), suppose $2^k\leqslant n<2^{k+1}$, $k\in\mathbf{N}^*$. Denote $A=\frac{1}{n}(x_1+\cdots+x_n)$. Then since the inequality holds for 2^{k+1}, we know that

$$f\left(\frac{x_1+\cdots+x_n+(2^{k+1}-n)A}{2^{k+1}}\right)\leqslant\frac{1}{2^{k+1}}\left(\sum_{j=1}^{n}f(x_j)+(2^{k+1}-n)f(A)\right).$$

At the same time,

$$\frac{1}{2^{k+1}}(x_1+\cdots+x_n+(2^{k+1}-n)A)=\frac{1}{2^{k+1}}(nA+(2^{k+1}-n)A)=A.$$

Then, we have

$$2^{k+1}f(A)\leqslant\sum_{j=1}^{n}f(x_j)+(2^{k+1}-n)f(A).$$

So $f(A)\leqslant\frac{1}{n}\sum_{j=1}^{n}f(x_j)$, i.e., the inequality holds for n.

The proposition is proved.

8. Lemma: Suppose $f(x)$ is a convex function on the interval $(0,1)$, $n\in\mathbf{N}^*$, $n\geqslant 2$. The positive real numbers $x_1,\cdots,x_n$ satisfy $x_1+\cdots+x_n=1$, then

$$\sum_{i=1}^{n}f(x_i)\geqslant\sum_{i=1}^{n}f\left(\frac{1-x_i}{n-1}\right).$$

Proof of the Lemma: From Jensen's Inequality, we know that

$$\sum_{i=1}^{n} f(x_i) = \sum_{i=1}^{n}\left(\frac{1}{n-1}\sum_{j\neq i} f(x_j)\right) \geqslant \sum_{i=1}^{n} f\left(\frac{1}{n-1}\sum_{j\neq i} x_j\right)$$
$$= \sum_{i=1}^{n} f\left(\frac{1-x_i}{n-1}\right).$$

So, the Lemma holds.

Let's go back to the original question. Let $f(x) = \ln\frac{1+x}{x}$. We notice that for any x, $y \in (0, 1)$, it is always true that

$$f(x)+f(y) = \ln\frac{1+x}{x} + \ln\frac{1+y}{y} = \ln\frac{1+xy+x+y}{xy}$$
$$= \ln\left(\frac{1}{xy} + \frac{x+y}{xy} + 1\right)$$
$$\geqslant \ln\left(\frac{1}{\left(\frac{x+y}{2}\right)^2} + \frac{x+y}{\left(\frac{x+y}{2}\right)^2} + 1\right)$$
$$= \ln\left(\frac{4}{(x+y)^2} + \frac{4}{x+y} + 1\right) = \ln\left(\frac{(x+y+2)^2}{(x+y)^2}\right)$$
$$= 2\ln\left(1 + \frac{1}{\frac{x+y}{2}}\right) = 2f\left(\frac{x+y}{2}\right).$$

So, $f(x) = \ln\frac{1+x}{x}$ is a convex function on $(0, 1)$. Considering this, together with previous conclusions, we may know that the proposition holds.

9. Denote $S_n = \sum_{i=1}^{n}\frac{F_i}{2^i}$, then $S_1 = \frac{1}{2}$, $S_2 = \frac{1}{2} + \frac{1}{4} = \frac{3}{4}$. While $n \geqslant 3$, it is true that

$$S_n = \frac{1}{2} + \frac{1}{4} + \sum_{i=3}^{n}\frac{F_i}{2^i}$$
$$= \frac{3}{4} + \sum_{i=3}^{n}\frac{F_{i-1}+F_{i-2}}{2^i}$$
$$= \frac{3}{4} + \frac{1}{2}\sum_{i=3}^{n}\frac{F_{i-1}}{2^{i-1}} + \frac{1}{4}\sum_{i=3}^{n}\frac{F_{i-2}}{2^{i-2}}$$

$$= \frac{3}{4} + \frac{1}{2}\sum_{i=2}^{n-1}\frac{F_i}{2^i} + \frac{1}{4}\sum_{i=1}^{n-2}\frac{F_i}{2^i}$$

$$= \frac{3}{4} + \frac{1}{2}\left(S_{n-1} - \frac{1}{2}\right) + \frac{1}{4}S_{n-2}$$

$$= \frac{1}{2} + \frac{1}{2}S_{n-1} + \frac{1}{4}S_{n-2}.$$

By use of $S_1 = \frac{1}{2}$ and $S_2 = \frac{3}{4}$, we may know that for $n = 1, 2$, it is always true $S_n < 2$. Now we suppose for $n = k, k+1$, it is true that $S_n < 2$, then we have

$$S_{k+2} = \frac{1}{2} + \frac{1}{2}S_{k+1} + \frac{1}{4}S_k < \frac{1}{2} + \frac{1}{2}\times 2 + \frac{1}{4}\times 2 = 2.$$

So, the proposition holds true.

10. By making use of $a_{n+2} = a_{n+1} + a_n$, we know that $a_9 = a_8 + a_7 = 2a_7 + a_6 = \cdots = 21a_2 + 13a_1$. According to the conditions set by the question, we can know that the following indeterminate equation has at least two sets of positive integer solutions (x, y), such that $x \leqslant y$:

$$13x + 21y = k. \qquad ①$$

We notice that if ① has two sets of positive integer solutions (x_1, y_1) and (x_2, y_2), such that $x_1 \leqslant y_1$, $x_2 \leqslant y_2$, then $13x_1 + 21y_1 = 13x_2 + 21y_2 = k$. By symmetry, without loss of generality, we may suppose $x_1 \leqslant x_2$. Then $13(x_2 - x_1) = 21(y_1 - y_2)$. Thus, by $x_1 = x_2$ we know that $y_1 = y_2$, which leads to $(x_1, y_1) = (x_2, y_2)$ and causes a contradiction. So $x_1 < x_2$. Therefore, we have $21 \mid x_2 - x_1$, $13 \mid y_1 - y_2$ (it makes use of $(13, 21) = 1$). So $x_2 - x_1 \geqslant 21$, and get $x_2 \geqslant 21 + x_1 \geqslant 22$. From $y_2 \geqslant x_2$, we know that $k \geqslant 13 \times 22 + 21 \times 22 = 748$.

On the other hand, when $k = 748$, ① has two sets of distinctive positive integer solutions, and they are $(22, 22)$ and $(1, 35)$, corresponding to (a_1, a_2) respectively, and we then have two sequences that meet all the requirements.

Generalizing all above, the least positive integer we look for is 748.

11. If $m \geqslant k+2$, then $F_m \geqslant F_{k+2} = F_{k+1} + F_k \geqslant 2F_k$ (since the sequence $\{F_n\}$ is a non-decreasing sequence). Then $x_1 \leqslant \frac{1}{2}$. From the definition of $\{x_n\}$ we may know that $x_2 \leqslant 0$, and further we may use mathematical induction to easily prove: for $n \geqslant 2$, it is always true that $x_n \leqslant 0$. At this time, the sequence $\{x_n\}$ does not contain any term equal to 1, so $m < k+2$. Since $m > k$, our conclusion has to be $m = k+1$.

On the other hand, for any $k \in \mathbf{N}^*$, if $m = k+1$, then from the definition of $\{x_n\}$, we can know that $x_2 = \frac{2F_k - F_{k+1}}{F_{k+1} - F_k}$ (unless $k=1$, $m=2$, then $x_1 = 1$, meaning there is a 1 in this sequence already). And we get $x_2 = \frac{F_k - F_{k-1}}{F_{k-1}} = \frac{F_{k-2}}{F_{k-1}}$. We conduct this reasoning recursively to find that: when k is an odd number, supposing $k = 2n+1$, we have $x_3 = \frac{F_{2n-3}}{F_{2n-2}}$, $\cdots$ $x_{n+1} = \frac{F_1}{F_2} = 1$. This is in accordance with the question; when k is an even number, supposing $k = 2n$, we have $x_3 = \frac{F_{2n-4}}{F_{2n-3}}$, $\cdots$, $x_n = \frac{F_2}{F_3} = \frac{1}{2}$, after which every term in the sequence is no bigger than 0, and this is not in accordance with the question.

Generalizing all above, the positive integer pairs we look for is $(k, m) = (2n-1, 2n)$, $n \in \mathbf{N}^*$.

12. The answer is 11 Yuan.

Suppose $f(n)$ is the smallest amount of money that needs to be paid to confirm the number Mr. Zhang takes from $\{1, 2, \cdots, n\}$, then $f(n)$ is a non-decreasing sequence. And, if the first subset chosen by Mr. Wang is a set with m elements, then

$$f(n) \leqslant \max\{f(m)+2,\ f(n-m)+1\}.$$

Next, we make use of the Fibonacci Sequence $\{F_n\}$ and prove the following conclusion: suppose x is a positive integer, and $F_n < x \leqslant F_{n+1}$ $(n \geqslant 2)$, then

$$f(x) = n - 1. \qquad ①$$

We first prove that for any $n \in \mathbf{N}^*$ $(n \geqslant 2)$, it is always true that

$$f(F_{n+1}) \leqslant n - 1. \quad ②$$

As a matter of fact, when $n = 2$, $F_3 = 2$, it is easy to know that $f(F_3) \leqslant 2$. We suppose for positive integers less than n, ② is always true. Then we consider the case with n. Mr. Wang for the first time takes one subset and makes its number of elements F_{n-1}, then it is true $f(F_{n+1}) \leqslant \max\{f(F_{n-1}) + 2, f(F_{n+1} - F_{n-1}) + 1\} = \max\{f(F_{n-1}) + 2, f(F_n) + 1\} \leqslant \max\{n - 1, n - 1\} = n - 1$ (here it is deemed that $f(F_2) = f(1) = 0$). So ② holds true for all positive integers n.

Next, we are to prove: for any $n \in \mathbf{N}^*$, $F_n < x \leqslant F_{n+1}$, $x \in \mathbf{N}^*$, it is always true that $f(x) \geqslant n - 1$.

When $n = 2$, $x = F_3 = 2$. Now, it is easy to know $f(2) \geqslant 2$, so the conclusion is true for $n = 1$. Suppose the proposition is true for positive integers less than n. We consider the case with n. For any $n \in \mathbf{N}^*$, $F_n < x \leqslant F_{n+1}$. (Attention, here we have $n \geqslant 3$, so $x \geqslant 3$.)

If the number of elements in the subset that Mr. Wang takes for the first time is $\leqslant F_{n-2}$, then the smallest amount of money that Mr. Wang needs to pay $\geqslant f(x - F_{n-2}) + 1 \geqslant f(F_{n-1} + 1) + 1 \geqslant n - 1$; if the number of elements in the subset that Mr. Wang takes for the first time is $\geqslant F_{n-2} + 1$, then the smallest amount of money that Mr. Wang needs to pay $\geqslant f(F_{n-2} + 1) + 2 \geqslant n - 3 + 2 = n - 1$. So, $f(x) \geqslant n - 1$.

Generalizing all above, the conclusion ① holds. Making use of this conclusion and concerning also $144 = F_{12}$, we may know that Mr. Wang at least has to pay 11 Yuan to guarantee getting to know the number Mr. Zhang takes.

13. When $m = 1$, it is obviously true. Now we consider the case $m \geqslant 2$.

We first prove that $\{F_n \pmod m\}$ is a periodic sequence. This can be observed by noticing that (F_n, F_{n+1}) has only m^2 kinds of different situations under $\bmod m$. By using the Drawer Principle (also known as the Pigeonhole Principle), we know there exists an $n < k$, such that

$(F_n, F_{n+1}) \equiv (F_k, F_{k+1}) \pmod{m}$, and then together with the recursive formula, we may deduce that $F_{n-1} \equiv F_{k-1} \pmod{m}$. Reversely deduce in turn and we can get that conclusion.

After that, from $F_1 = F_2 \equiv 1 \pmod{m}$, we know that there exists a $p \in \mathbf{N}^*$, such that $F_{p+1} \equiv F_{p+2} \equiv 1 \pmod{m}$, hence $F_p \equiv 0 \pmod{m}$, $F_{p-1} \equiv 1 \pmod{m}$, $F_{p-2} \equiv -1 \pmod{m}$, and further, for $t \in \mathbf{N}^*$, we have $F_{tp-2} \equiv -1 \pmod{m}$. Take $n = tp - 2$, then we have $F_n^4 - F_n - 2 \equiv 1 + 1 - 2 \equiv 0 \pmod{m}$. The proposition is then proved.

14. (1) We can't. As a matter of fact, if we can partition $\mathbf{N}^*$ into the union of m F-sequences, then we consider the positive integers: $2m$, $2m + 1$, $\cdots$, $4m$. Among all these $2m + 1$ numbers, there must be 3 numbers that are from the same F-sequence. However, take any three numbers from the $2m + 1$ numbers, and the sum of any two of them is bigger than the third number. This becomes a contradiction.

(2) We use the Fibonacci expression (see Example 2 in Section 9) for positive integers to prove: $\mathbf{N}^*$ could be partitioned into the union of infinitely many F-sequences.

We will, under Fibonacci expression, make all positive integers that make $a_2 = 1$ arranged, from the smallest to biggest, on the first row; make all positive integers that make $a_2 = 0$, while $a_3 = 1$ arranged, from the smallest to biggest, on the second row; make all positive integers that make $a_2 = a_3 = 0$ while $a_4 = 1$ arranged, from the smallest to biggest, on the third row; ... and a table is listed below:

F_2	$F_2 + F_4$	$F_2 + F_5$	$F_2 + F_6$	$F_2 + F_4 + F_6$	$\cdots$
F_3	$F_3 + F_5$	$F_3 + F_6$	$F_3 + F_7$	$F_3 + F_5 + F_7$	$\cdots$
F_4	$F_4 + F_6$	$F_4 + F_7$	$\cdots$	$\cdots$	$\cdots$
$\cdots$	$\cdots$	$\cdots$	$\cdots$	$\cdots$	$\cdots$

By Zeckendorf's Theorem, we know that every positive integer appears exactly once on the above table, while every column from top to bottom forms a F-sequence. So, the conclusion of (2) is surely true.

15. When $i=1$, it is obviously true that $a_1 \leqslant k$.

Suppose for $1 \leqslant s < n$, we have $sa_s \leqslant k$. Next, we are to prove: $(s+1)a_{s+1} \leqslant k$.

If $(s+1)a_{s+1} \leqslant sa_s$, then of course we have $(s+1)a_{s+1} \leqslant k$; if $(s+1)a_{s+1} > sa_s$, i.e., $a_{s+1} > s(a_s - a_{s+1})$, then $\frac{a_{s+1}}{a_s - a_{s+1}} > s$.

Noticing that $[a_s, a_{s+1}] = \frac{a_s a_{s+1}}{(a_s, a_{s+1})}$, making use of $\frac{a_{s+1}}{(a_s, a_{s+1})} \geqslant \frac{a_{s+1}}{a_s - a_{s+1}} > s$, and concerning also $\frac{a_{s+1}}{(a_s, a_{s+1})} \in \mathbf{N}^*$, we may know that $\frac{a_{s+1}}{(a_s, a_{s+1})} \geqslant s+1$. Then

$$(s+1)a_{s+1} < (s+1)a_s \leqslant \frac{a_{s+1}}{(a_s, a_{s+1})} \cdot a_s = [a_s, a_{s+1}] \leqslant k.$$

So the proposition is also true for $s+1$.

16. We use mathematical induction (inducting with n) to prove the following strengthened proposition:

$$\frac{1}{[a_0, a_1]} + \frac{1}{[a_1, a_2]} + \cdots + \frac{1}{[a_{n-1}, a_n]} \leqslant \frac{1}{a_0}\left(1 - \frac{1}{2^n}\right). \quad ①$$

When $n=1$, from the condition $a_0 < a_1$, we know that $[a_0, a_1] \geqslant 2a_0$, then $\frac{1}{[a_0, a_1]} \leqslant \frac{1}{2a_0} = \frac{1}{a_0}\left(1 - \frac{1}{2}\right)$. So when $n=1$, the inequality ① holds.

Suppose that the inequality ① holds true for n, then for the case $n+1$, we have

$$\begin{aligned} &\frac{1}{[a_0, a_1]} + \frac{1}{[a_1, a_2]} + \cdots + \frac{1}{[a_n, a_{n+1}]} \\ \leqslant\ &\frac{1}{[a_0, a_1]} + \frac{1}{a_1}\left(1 - \frac{1}{2^n}\right). \end{aligned} \quad ②$$

If $a_1 \geqslant 2a_0$, then the right-hand side of ② $\leqslant \frac{1}{[a_0, a_1]} + \frac{1}{2a_0}\left(1 - \frac{1}{2^n}\right) \leqslant \frac{1}{2a_0}\left(2 - \frac{1}{2^n}\right) = \frac{1}{a_0}\left(1 - \frac{1}{2^{n+1}}\right)$; if $a_0 < a_1 < 2a_0$, then

from $(a_0, a_1) \leqslant a_1 - a_0$, we know that the right-hand side of ②
$\leqslant \frac{a_1 - a_0}{a_0 a_1} + \frac{1}{a_1}\left(1 - \frac{1}{2^n}\right) = \frac{1}{a_0} - \frac{1}{a_1 \cdot 2^n} < \frac{1}{a_0} - \frac{1}{2a_0 \cdot 2^n} = \frac{1}{a_0}\left(1 - \frac{1}{2^{n+1}}\right)$.
So, the inequality ① is also true for $n+1$.

17. Suppose under base-2, $n = (a_k a_{k-1} \cdots a_0)_2$, where $a_i \in \{0, 1\}$, $a_k = 1$. Let $t_n = (a_k a_{k-1} \cdots a_0)_3$ (which is a positive integer expressed under base-3), $t_0 = 0$.

We use mathematical induction to prove: for any $n \in \mathbf{N}^*$, it is always true that $u_n = t_n$.

When $n = 1$, the proposition is obviously true. Suppose for any $m < n$, it is always true that $u_m = t_m$. Next, we are to prove $u_n = t_n$.

On the one hand, no any three numbers in the sequence $\{t_0, t_1, \cdots, t_n\}$ form an arithmetic sequence. This is because for any $0 \leqslant \alpha < \beta < \gamma \leqslant n$, if $t_\alpha + t_\gamma = 2t_\beta$, then since $2t_\beta$ under base-3 is composed only by numbers 0 and 2, we know that every corresponding numbers of t_α, t_γ, under base-3 are exactly the same. Therefore, $t_\alpha = t_\gamma$, and this requires that $\alpha = \gamma$. This brings up a contradiction.

The above discussions show that $u_n \leqslant t_n$.

On the other hand, if $u_n < t_n$, then from induction hypothesis we know that $u_n \in \{t_{n-1} + 1, \cdots, t_n - 1\}$. At this time, under the ternary expression for u_n, the number 2 will necessarily appear (since the positive integer under base-3 with only numbers 0 and $1 \in \{t_0, t_1, \cdots\}$). So, there exist $a, b \in \mathbf{N}$ such that $0 \leqslant t_a < t_b < u_n$ satisfying:

(1) If within the ternary expression of u_n, a certain digit is a 0 (or 1), then for the ternary expression of t_a, t_b, the corresponding digits will also show a 0 (or 1) on the same position.

(2) If within the ternary expression of u_n, a certain digit shows a 2, then on the same position of t_a, it should show a 0, while t_b shows a 1 on the same position.

Then $t_a + u_n = 2t_b$. This is a contradiction. So $u_n \geqslant t_n$.

Generalizing all above, we may know that $u_n = t_n$. Considering $100 = (1\,100\,100)_2$, we have $u_{100} = (1\,100\,100)_3 = 981$.

18. We will discuss this problem under base-b.

Suppose among all the numbers that are divisible by $b^n - 1$, the smallest value of the number of all non-zero digits under base-b is s. Among all the numbers whose total numbers of all non-zero digits are s, we take the number A with the smallest sum of all digits.

Suppose $A = a_1 b^{n_1} + a_2 b^{n_2} + \cdots + a_s b^{n_s}$ is the base-b expression for A, where $n_1 > n_2 > \cdots > n_s \geqslant 0$, $1 \leqslant a_i < b$, $i = 1, 2, \cdots, s$.

Next, we prove: $n_1, n_2, \cdots, n_s$ constitute a complete system with modulus n, therefore $s \geqslant n$.

On the one hand, suppose $1 \leqslant i < j \leqslant s$. If $n_i \equiv n_j \equiv r \pmod n$, where $0 \leqslant r \leqslant n-1$. We examine the numbers

$$B = A - a_i b^{n_i} - a_j b^{n_j} + (a_i + a_j) b^{nn_1 + r}.$$

Obviously $b^n - 1 \mid B$. If $a_i + a_j < b$, then the total number of all non-zero digits of B is $s - 1$, which is contradictory to the choice for A. So, it must be $b \leqslant a_i + a_j < 2b$. Suppose $a_i + a_j = b + q$, $0 \leqslant q < b$, then at this time, the base-b expression of B is

$$\begin{aligned} B &= b^{nn_1 + r + 1} + q b^{nn_1 + r} + a b_1^{n_1} + \cdots + a_{i-1} b^{n_{i-1}} + a_{i+1} b^{n_{i+1}} \\ &\quad + \cdots + a_{j-1} b^{n_j - 1} + a_{j+1} b^{n_{j+1}} + \cdots + a_s b^{n_s}. \end{aligned}$$

Then, the sum of all digits of

$$B = \sum_{k=1}^{s} a_k - (a_i + a_j) + 1 + q = \sum_{k=1}^{s} a_k + 1 - b < \sum_{k=1}^{s} a_k,$$

which is contradictory to the choice for A. So, any two from $n_1, \cdots, n_s$ are not congruent under mod n.

On the other hand, if $s < n$, then we suppose $n_i \equiv r_i \pmod n$, $0 \leqslant r_i < n$. We examine the number C, where

$$C = a_1 b^{r_1} + a_2 b^{r_2} + \cdots + a_s b^{r_s}.$$

Due to $b^{n_i} \equiv b^{r_i} \pmod{b^n - 1}$, we have $b^n - 1 \mid C$. But $s < n$ means

$$0 < C \leqslant (b-1)b + (b-1)b^2 + \cdots + (b-1)b^{n-1} = b^n - b < b^n - 1.$$

This is a contradiction.

So, the proposition holds.

19. We start from every odd prime number to find n that satisfies the condition.

We notice that for any $m \in \mathbf{N}^*$, $(p^{2^m}-1, p^{2^m}+1)=2$. So, from the formula for the difference of two squares, considering together with mathematical induction, we may prove that any two numbers from $p+1$, p^2+1, $\cdots$, $p^{2^m}+1$ have no same odd and prime factors, and none of these numbers is a multiple of p. So, there exists an $m \in \mathbf{N}^*$, such that

$$h(p^{2^{m-1}}+1) < p < h(p^{2^m}+1). \qquad ①$$

Take the smallest positive integer m_0 that satisfies ①. Let $n = p^{2^{m_0}} - 1$. We assert that $h(n) < h(n+1) < h(n+2)$.

As a matter of fact,

$$n = p^{2^{m_0}} - 1 = (p-1)(p+1)(p^2+1)\cdots(p^{2^{m_0}}+1),$$

while m_0 is the smallest positive integer that satisfies ①, so we have $h(n) < p = h(n+1)$. And since $h(n+2) = h(p^{2^{m_0}}+1)$, from ① we know that $h(n+1) < h(n+2)$.

The above discussion shows: for every odd and prime number p, there is always an n that meets the conditions (obviously different p's correspond to different n's). And since there are infinitely many odd and prime numbers, the number of n that meet all the conditions should be infinite.

20. Lemma: If $k \in \mathbf{N}^*$, $k \neq 3$ and k is not a power of 2, then

$$w(2^k+1) > 1.$$

As a matter of fact, if $2^k+1 = p^m$, where p is a prime number, and $m \in \mathbf{N}^*$, we denote $k = 2^\alpha \cdot \beta$, $\alpha \geqslant 0$, $\beta > 1$, and β is an odd number. We discuss by two cases:

(1) $\alpha = 0$, then from $k \neq 3$ we know that $\beta > 3$. So, $2^\beta + 1 = (2+1)(2^{\beta-1} - 2^{\beta-2} + \cdots + 1)$ is a multiple of 3, and $2^\beta + 1 > 9$. If $2^\beta + 1 = 3^\gamma$, then $\gamma \geqslant 3$. At this time, we take mod 4 on both sides, then we know that $(-1)^\gamma \equiv 1 \pmod 4$, so γ is an even number. Denote $\gamma = 2\delta$,

then $2^{\beta} = (3^{\delta} - 1)(3^{\delta} + 1)$. Since $3^{\delta} - 1$ and $3^{\delta} + 1$ are adjacent even numbers and their product is a power of 2, then we must have $3^{\delta} - 1 = 2$, which leads to $\delta = 1$, $\gamma = 2$. This is a contradiction. So, when $\alpha = 0$, the lemma holds true.

(2) $\alpha > 0$. At this time, by making use of factoring we can know that $2^{2^{\alpha}} + 1 \mid 2^k + 1$. If $w(2^k + 1) = 1$, then $p = 2^{2^{\alpha}} + 1$ is a prime number. At this time, we suppose $2^{2^{\alpha} \cdot \beta} + 1 = p^u$, i.e., $(p - 1)^{\beta} + 1 = p^u$, $u \geqslant 2$. We take mod p^2 on both sides and make use of the Binomial Theorem, then we know that $p \mid \beta$. Further, suppose $\beta = p^v \cdot x$, $p \nmid x$, and we can know from the Binomial Theorem that

$$p^u = p^{\beta} - C_{\beta}^{\beta-1} p^{\beta-1} + \cdots + C_{\beta}^2 p^2 - \beta \cdot p.$$

The last term on the right is a multiple of p^{v+1}, but not a multiple of p^{v+2}, while every other term is a multiple of p^{v+2}. So, the above equation is not true. So, when $\alpha > 0$, the lemma is also true.

Through the above lemma, we know that when $k \neq 3$ and k is not a power of 2, we have $w(2^k) < w(2^k + 1)$. Next, we are to prove that there exist infinitely many such k, such that $w(2^k + 1) < w(2^k + 2)$.

As a matter of fact, if we have only a limited number of k as above such that $w(2^k + 1) < w(2^k + 2)$, then there exists $k_0 = 2^q > 5$. For every $k \in \{k_0 + 1, \cdots, 2k_0 - 1\}$, it is always true that $w(2^k + 1) \geqslant w(2^k + 2) = 1 + w(2^{k-1} + 1)$. Then, we have

$$w(2^{2k_0 - 1} + 1) \geqslant 1 + w(2^{2k_0 - 2} + 1) \geqslant \cdots \geqslant (k_0 - 1) + w(2^{k_0} + 1) \geqslant k_0.$$

This requires that $2^{2k_0 - 1} + 1 \geqslant p_1 \cdots p_{k_0}$, where $p_1, \cdots, p_{k_0}$ are the initial k_0 prime numbers. However,

$$p_1 \cdots p_{k_0} \geqslant (2 \times 3 \times 5 \times 7 \times 11) \times (p_6 \cdots p_{k_0}) > 4^5 \cdot 4^{k_0 - 5} = 2^{2k_0}.$$

Thus we have a contradiction. So, the proposition holds true.

21. Suppose the permutation of prime numbers from the smallest to biggest is $p_1, p_2, \cdots, p_n, \cdots$. Then $a_n = p_1 + p_2 + \cdots + p_n$.

When $n = 1, 2, 3, 4$, we can then directly verify the proposition and then know that it holds true. Now we suppose that the proposition holds true with $n - 1$, i.e., there exists a positive integer x, such that

$a_{n-1} \leqslant x^2 \leqslant a_n$. We take the biggest x which meets the requirement and denote it to be y, then $y^2 \leqslant a_n$, while $(y+1)^2 > a_n$, where $n \geqslant 5$.

Denote $p_{n+1} = 2k+1$. Then when $n \geqslant 5$, concerning that two adjacent prime numbers differ by at least 2, we know

$$p_1 + p_2 + \cdots + p_n < 1 + 3 + 5 + \cdots + (2k-1) = k^2.$$

Hence, $y^2 \leqslant a_n < k^2$, i.e., $y < k$. Then

$$\begin{aligned}(y+1)^2 &= y^2 + 2y + 1 < y^2 + 2k + 1 = y^2 + p_{n+1} \\ &\leqslant p_1 + \cdots + p_n + p_{n+1} = a_{n+1}.\end{aligned}$$

So, the proposition also holds true for n.

From all of above, for all $n \in \mathbf{N}^*$, the proposition holds true.

22. We suppose a to be a positive odd number. If $(a, 5) = 1$, then $(a, 10) = 1$. In the sequence 1, 11, $\cdots$, $\underbrace{11\cdots1}_{a\ 1\text{'s}}$, there are two numbers being congruent under $\bmod\ a$, i.e., there exist $1 \leqslant i < j \leqslant a$, such that $\underbrace{11\cdots1}_{j\ 1\text{'s}} \equiv \underbrace{1\cdots1}_{i\ 1\text{'s}} \pmod a$. Namely, $a \mid \underbrace{1\cdots1}_{j-i\ 1\text{'s}}\ \underbrace{0\cdots0}_{i\ 0\text{'s}}$, so $a \mid \underbrace{1\cdots1}_{j-i\ 1\text{'s}}$. The proposition is then proved.

If $5 \mid a$, then suppose $a = 5^\alpha \cdot b$, $\alpha \in \mathbf{N}^*$, $(5, b) = 1$. We prove the following lemma in the first place.

Lemma: For any positive integer n, there exists an n-digit positive integer A_n with 1, 3, 5, 7, 9 as digits only, such that $5^n \mid A_n$.

We prove this lemma by using induction. When $n = 1$, just take $A_n = 5$. Suppose when $n = k$, there exists an k-digit number A_k, whose digits belong to the set $\{1, 3, 5, 7, 9\}$, and $5^k \mid A_k$. Consider the following numbers

$$\begin{gathered}10^k + A_k,\ 3 \times 10^k + A_k,\ 5 \times 10^k + A_k, \\ 7 \times 10^k + A_k,\ 9 \times 10^k + A_k.\end{gathered}$$

If $5^{k+1} \mid A_k$, then it suffices to let $A_{k+1} = 5 \times 10^k + A_k$; if $5^{k+1} \nmid a_k$, then suppose $a_k = 5^k \times t$, where $t = r \pmod 5$, $r \in \{1, 2, 3, 4\}$. Noticing that $(5, 2^k) = 1$, then we know $\{2^k, 3 \times 2^k, 7 \times 2^k, 9 \times 2^k\}$ becomes a reduced residue system with modulus 5. Then, we may choose $S \in \{1, 3, 7, 9\}$ such that $S \times 2^k \equiv 5 - r \pmod 5$. Therefore, by letting $A_{k+1} = S \times 10^k + A_k$, we have $5^{k+1} \mid A_{k+1}$, and all digits of A_{k+1} belong to $\{1, 3, 5, 7, 9\}$. The lemma is proved.

Going back to the original question. From lemma, we can know that there exists an α-digit number A, such that $5^\alpha \mid A$. Then, within the sequence $\overline{A}$, $\overline{AA}$, $\cdots$, $\underbrace{\overline{A\cdots A}}_{b\ A\text{'s}}$ (here $\underbrace{\overline{A\cdots A}}_{k\ A\text{'s}}$ means the positive integer we get by writing consecutively the k A's), there must be two congruent under mod b. By making use of the method in the first case we know that the proposition holds true.

23. (1) Let $x_1 = 123\,467\,895$, then $S(x_1) = 45$, and from $45 \mid 123\,467\,895$ we may know that $x_1 \in A$. Now we suppose $x_k \in A$, and the expression of x_k, under the decimal system, has equal numbers of appearances for the numbers 1, 2, $\cdots$, 9. We suppose that x_k is an m-digit number, and we take $x_{k+1} = x_k \cdot (10^{2m} + 10^m + 1) = \overline{x_k x_k x_k}$, then under the decimal system, the numbers of appearances for the numbers 1, 2, $\cdots$, 9 are the same in x_{k+1}, and $S(x_{k+1}) = 3S(x_k)$. Since $10^{2m} + 10^m + 1 \equiv 1 + 1 + 1 \equiv 0 \pmod 3$, and also $S(x_k) \mid x_k$ we know that $S(x_{k+1}) \mid x_{k+1}$. By this, together with mathematical induction, we know that the conclusion (1) holds true.

(2) **Lemma:** For any $n \in \mathbf{N}^*$, there exists an n-digit positive integer x_n, whose digits are 1 and 2 only, such that $2^n \mid x_n$.

We can follow the proof for the lemma in the previous question to get this lemma proved.

Going back to the original question. When $k = 1, 2, 3, 4, 5$, we take 1, 12, 112, 4112, and 42 112, and we see that the proposition holds true.

When $k \geqslant 6$, then our idea is to look for a k-digit number x in A: we will look for x whose last n digits is the x_n mentioned in the lemma, and then we fill in non-zero numbers in front of x_n, and the sum of all digits from the k-digit positive integer x formed is 2^n, where n is to be determined.

A sufficient condition for the above-mentioned n to exist is:

$$S(x_n)+(k-n)\leqslant 2^n\leqslant S(x_n)+9(k-n). \quad ①$$

Since $n\leqslant S(x_n)\leqslant 2n$, therefore, if the following is satisfied, then ① holds.

$$2n+(k-n)\leqslant 2^n\leqslant n+9(k-n),$$

namely,

$$n+k\leqslant 2^n\leqslant 9k-8n. \quad ②$$

Next, we prove: when $k\geqslant 6$, there exists an $n\in\mathbf{N}^*$ that satisfies ②.

As a matter of fact, suppose n is the biggest positive integer satisfying $2^n+8n\leqslant 9k$, then $9k<2^{n+1}+8(n+1)$. This suggests that if $2^{n+1}+8(n+1)\leqslant 9(2^n-n)$, then n satisfies ②.

We note that $k\geqslant 6$, so the above-mentioned n satisfies $n\geqslant 4$. Then $7\times 2^n\geqslant 17n+8$ (this inequality can be proved through inducting with n), and this means $2^{n+1}+8(n+1)\leqslant 9(2^n-n)$. Therefore n satisfies ②.

Generalizing all above, we know that the conclusion ② is true.

24. There exists a sequence that satisfies the conditions.

We arrange all prime numbers bigger than 5 from the smallest to biggest, and we then get the sequence $p_0, p_1, \cdots$; we define a sequence $\{q_n\}$ as follows: $q_{3k}=6$, $q_{3k+1}=10$, $q_{3k+2}=15$, $k=0,1,2,\cdots$. Now we define the sequence $\{a_n\}$ as $a_n=p_nq_n$, $n=0,1,2,\cdots$. We will prove that the sequence $\{a_n\}$ meets all the conditions.

We notice that for subscript $i\neq j$, we have $p_i\neq p_j$, so there is no term in $\{a_n\}$ that is a multiple of any other term. Hence, (1) is satisfied. We now take a step further. If $i\equiv j \pmod 3$, then $(a_i, a_j)=(q_i, q_j)=6$, 10, or 15; if $i\not\equiv j \pmod 3$, then since 6, 10, 15 are mutually non-prime for any two of them, we know that $(a_i, a_j)=(q_i, q_j)>1$. Moreover, $5\nmid a_0$, $3\nmid a_1$, $2\nmid a_3$, and every prime number bigger than 5 divides at most one term from the sequence $\{a_n\}$. Hence, there is no positive integer bigger than 1 that divides every term from $\{a_n\}$. So (2) is also satisfied.

25. We prove: when $k=2,\cdots,p-1$, there exists a set $\{b_{k,1},$

$b_{k,2}, \cdots, b_{k,k}\}$, where $b_{k,i} = 1$, or the product of some numbers from $a_1, \cdots, a_{k-1}$ satisfying that for $1 \leqslant i < j \leqslant k$, it is always true that $b_{k,i} \not\equiv b_{k,j} \pmod p$.

When $k = 2$, from the conditions, we know that $a_1 \not\equiv 1 \pmod p$. Then just take $\{b_{k,1}, b_{k,2}\} = \{1, a_1\}$. Now we suppose ① is true for $k(2 \leqslant k \leqslant p-2)$. From the condition $p \nmid a_k$ we know that $a_k b_{k,1}, \cdots, a_k b_{k,k}$ are not congruent for any two of them under mod p. By the composition (none of the terms is a multiple of p) of the sequence $\{b_{k,1}, \cdots, b_{k,k}\}$, and that $a_k^k \not\equiv 1 \pmod p$, we know

$$(a_k b_{k,1})\cdots(a_k b_{k,k}) \not\equiv b_{k,1}\cdots b_{k,k} \pmod p.$$

So, under mod p, $(a_k b_{k,1}, \cdots, a_k b_{k,k})$ is not a permutation of $(b_{k,1}, \cdots, b_{k,k})$, therefore, there exists a $j \in \{1, 2, \cdots, k\}$ such that any two numbers from the sequence $\{a_k b_{k,j}, b_{k,1}, \cdots, b_{k,k}\}$ are not congruent under mod p. By this, we can know that the conclusion ① holds.

Once we examine $\{b_{p-1,1}, \cdots, b_{p-1,p-1}\}$ we can get the conclusion required by the question (because they constitute the reduced residue system with mod p).

26. (1) Take any $a \in \mathbf{N}^*$. Since there is a limited number of values of f that are $\leqslant f(a)$, so there exists an $n \in \mathbf{N}^*$, such that for $d \geqslant n$, we always have $f(a) < f(a+d)$. We consider the sequence $f(a), f(a+n), f(a+2n), \cdots, f(a+2^k n), f(a+2^{k+1}n), \cdots$.

If there exists a $k \in \mathbf{N}$ such that $f(a+2^{k+1}n) > f(a+2^k n)$, then we take $d = 2^k n$, and we know that (1) holds. So, for $k \in \mathbf{N}$, we have $f(a+2^{k+1}n) < f(a+2^k n)$ (here we do not take "equal sign", since f is an injection), namely, $f(a+n) > f(a+2n) > \cdots$. But since it is a surjection, there is only a limited number of values of f that are less than that of $f(a+n)$. This brings up a contradiction.

(2) They do not necessarily exist. For example, let $f: \mathbf{N}^* \to \mathbf{N}^*$ be defined below:

$$n = 1;\ 2;\ 3,\ 4;\ 5,\ 6,\ 7,\ 8;\ 9,\ 10,\ \cdots$$
$$f(n) = 1;\ 2;\ 4,\ 3;\ 8,\ 7,\ 6,\ 5;\ 16,\ 15,\ \cdots$$

In the definition above, for $n \in \mathbf{N}^*$, we have $f(2^n+1)=2^{n+1}$, $f(2^n+2)=2^{n+1}-1$, $\cdots$, $f(2^{n+1})=2^n+1$, while $f(1)=1$, $f(2)=2$.

Next, we prove: when $m \geqslant 5$, for a, $d \in \mathbf{N}^*$, it is always true that $f(a+(m-2)d) > f(a+(m-1)d)$, or $f(a+(m-1)d) > f(a+md)$.

As a matter of fact, if not true, then

$$f(a+(m-2)d) < f(a+(m-1)d) < f(a+md).$$

From the definition of f, we know that $f(a+(m-2)d)$, $f(a+(m-1)d)$, $f(a+md)$ are separately located on three different decreasing intervals, while the length of the decreasing interval from 2^n+1 to 2^{n+1} is 2^n. So, the length of the decreasing interval where $a+(m-1)d$ is located is $\geqslant \frac{a+(m-1)d}{2}$. Since $a+(m-2)d$ and $a+md$ are neither located on the decreasing interval where $a+(m-1)d$ is located, so $a+md-(a+(m-2)d) \geqslant \frac{a+(m-1)d}{2}$, leading to $4d \geqslant a+(m-1)d \geqslant a+4d$. This brings up a contradiction. Hence the conclusion of (2) is that they do not necessarily exist.

27. Lemma: If the integers n_1, n_2, $\cdots$ satisfy $n_{k+1} \geqslant n_k^2$, $k=1, 2, \cdots$, $n_1 > 1$, then for any $j \in \mathbf{N}^*$, we have

$$\prod_{k=j}^{+\infty}\left(1+\frac{1}{n_k}\right) \in \left(1+\frac{1}{n_j},\ 1+\frac{1}{n_j-1}\right].$$

Proof of the lemma: From the conditions, we know that

$$\prod_{k=j}^{+\infty}\left(1+\frac{1}{n_k}\right) \leqslant \prod_{k=0}^{+\infty}\left(1+\frac{1}{n_j^{2^k}}\right) = \prod_{k=0}^{+\infty}\left(1+\left(\frac{1}{n_j}\right)^{2^k}\right)$$

$$=\sum_{k=0}^{+\infty}\left(\frac{1}{n_j}\right)^k = \frac{1}{1-\frac{1}{n_j}} = 1+\frac{1}{n_j-1}.$$

So, the lemma holds.

From the lemma, we can prove uniqueness. (As a matter of fact, if

$$\alpha = \prod_{k=1}^{+\infty}\left(1+\frac{1}{n_k}\right) = \prod_{k=1}^{+\infty}\left(1+\frac{1}{m_k}\right),$$

and $n_1 = m_1$, $\cdots$, $n_j = m_j$, then

$$\alpha \Big/ \prod_{k=1}^{j}\left(1+\frac{1}{n_k}\right) = \alpha \Big/ \prod_{k=1}^{j}\left(1+\frac{1}{m_k}\right).$$

By the lemma, the former one $\in \left(1+\frac{1}{n_{j+1}},\ 1+\frac{1}{n_{j+1}-1}\right]$, and the latter one $\in \left(1+\frac{1}{m_{j+1}},\ 1+\frac{1}{m_{j+1}-1}\right]$. Then we can come to $n_{j+1} = m_{j+1}$.)

The existence could be obtained by the following method. Denote $\alpha_1 = \alpha \in (1, 2]$, then there exists a unique $n_1 \in \mathbf{N}^*$, such that $\alpha_1 \in \left(1+\frac{1}{n_1},\ 1+\frac{1}{n_1-1}\right]$. Let $\alpha_2 = \frac{\alpha_1}{1+\frac{1}{n_1}}$, then $1 < \alpha_2 < \alpha_1 \leqslant 2$. For this α_2, there exists a unique n_2, such that $\alpha_2 \in \left(1+\frac{1}{n_2},\ 1+\frac{1}{n_2-1}\right]$. Conducting the reasoning in turn, we may define a sequence $\{n_k\}_{k=1}^{+\infty}$. Next, we prove $n_k^2 \leqslant n_{k+1}$.

As a matter of fact,

$$1+\frac{1}{n_{k+1}} < \alpha_{k+1} = \frac{\alpha_k}{1+\frac{1}{n_k}} \leqslant \frac{1+\frac{1}{n_k-1}}{1+\frac{1}{n_k}} = 1+\frac{1}{n_k^2-1},$$

so $n_k^2 \leqslant n_{k+1}$.

Finally, from the definition of n_k we may know that

$$1 < \frac{\alpha}{\prod_{k=1}^{N}\left(1+\frac{1}{n_k}\right)} = \prod_{k=N}^{+\infty}\left(1+\frac{1}{n_k}\right) \leqslant 1+\frac{1}{n_{N+1}-1}.$$

Let $N \to +\infty$, and then we may get $\alpha = \prod_{k=1}^{+\infty}\left(1+\frac{1}{n_k}\right)$.

28. We firstly prove that there exists a pair of positive integers (p, q), such that $p < q$, and $a_p \mid a_q$. We examine the number table below:

$$x_{0,1}, x_{0,2}, \cdots, x_{0,m}$$
$$x_{1,1}, x_{1,2}, \cdots, x_{1,m}$$
$$\cdots$$
$$x_{m,1}, x_{m,2}, \cdots, x_{m,m}$$

Here $x_{0,1} = a_1$, $x_{0,j} = x_{0,j-1} + 1$, $j = 2, \cdots, m$. And,

$$x_{i,j} = \left(\prod_{k=1}^{m} x_{i-1,k}\right) + x_{i-1,j}.\ 1 \leqslant i, j \leqslant m.$$

In the above table, every row contains consecutive m positive integers, and for any two numbers a, b in each column, if $a < b$, then $a \mid b$.

By the conditions, every row contains at least two numbers from $\{a_n\}$. Hence, there are at least $2(m+1)$ numbers in the above table that are terms from the sequence $\{a_n\}$ and therefore, there are two numbers from one column in the table that are from the sequence $\{a_n\}$, and we denote them to be a_p, a_q, $p < q$, then we have $a_p \mid a_q$.

Now we give the value $a_q + 1$ to $x_{0,1}$, and by similar methods as above, we may construct a table of the same property. Then we can find the next pair (p', q'), $p' < q'$, such that $a_{p'} \mid a_{q'}$. Conduct reasoning like this, and we can find infinitely many pairs of (p, q), such that $p < q$, and $a_p \mid a_q$.

The proposition is proved.

29. This kind of partition exists. Let $A = \{n \in \mathbf{N} \mid$ The binary expression of n has an even number times of appearances of number $1\} = \{0, 3, 5, 6, \cdots\}$; $B = \{n \in \mathbf{N} \mid$ The binary expression of n has an odd number times of appearances of number $1\} = \{1, 2, 4, 7, \cdots\}$. We say that A, B are partitions that meet the conditions.

Next, we prove: for any $n \in \mathbf{N}$, it is always true that

$$r_A(n) = r_B(n). \qquad ①$$

We prove it through inducting with respect to m, the number of digits for n under binary expression.

When $m = 0, 1$, we notice that $r_A(0) = r_B(0) = r_A(1) = r_B(1) = 0$, and we know that ① holds.

Now we suppose ① holds for $n \in \mathbf{N}$, whose number of digits is no bigger than m. Let us examine the positive integer n with $m+1$ digits. For possible equalities $n = s_1 + s_2$, $s_1 > s_2$, $s_1, s_2 \in A$ (we do similar discussions when $s_1, s_2 \in B$), we will discuss with three cases.

Case 1. If the $m+1$th digit, from right to left, of s_1 is 1, then the $m+1$th digit, from right to left, of s_2 must be 0. Examine the two numbers from the 1st digit to the mth digit, from right to left. Among them, there is an odd number of 1 in s_1, and there is an even number of 1 in s_2. Let $s_1' = s_2 + 2^m$, $s_2' = s_1 - 2^m$, then both s_1' and s_2' have odd number of 1, and $s_1' > s_2'$, $s_1' + s_2' = n$, $s_1', s_2' \in B$. Conversely, when $s_1, s_2 \in B$, we also have $s_1', s_2' \in A$. So the numbers of expressions for this part in two sets are the same.

Case 2. If the $m+1$ th digits, from right to left, of s_1 and s_2 are both 0, and the mth digits are both 1, from right to left, then similar to above discussions, we may know that $s_1' = s_1 - 2^{m-1} \in B$, $s_2' = s_2 - 2^{m-1} \in B$. So (s_1', s_2') constitutes an expression of $n - 2^m$ in B. Conversely, when $s_1, s_2 \in B$, (s_1', s_2') constitutes an expression of $n - 2^m$ in A. By making use of $r_A(n - 2^m) = r_B(n - 2^m)$ (induction hypothesis), we may know that the numbers of expressions for this part in two sets are the same.

Case 3. If the $m+1$ th digits, from right to left, of s_1 and s_2 are both 0, and the mth digits are not both 1, from right to left, then the mth digit of s_1, from right to left, is 1, and the mth digit of s_2, from right to left, is 0. At this time, we examine the numbers of 1s for these two numbers from the 1st digit to the $m-1$ th digit, from right to left. Then we know that there is an odd number of 1s in s_1, and an even number of 1s in s_2. Let $s_1' = s_2 + 2^{m-1}$, $s_2' = s_1 - 2^{m-1}$. Similar to Case 1, we may know that the numbers of expressions for this part in two sets are the same.

Generalizing all above, for n with $m+1$ digits, we also have $r_A(n) = r_B(n)$. So, A, B defined above meet the conditions.

30. Suppose that S is the set of all the numbers that can be expressed in accordance with the conditions set in the question. Let $T = S \cup \{1\}$, and denote the set comprised of all the elements in S that are powers of 3 and are no bigger than h to be S_h, and $T_h = S_h \cup \{1\}$. Then the following conclusion is obviously true.

(1) $2T \subseteq T$, $3T \subseteq T$. Here $xT = \{xt \mid t \in T\}$.

(2) If $h < k$, then $2T_h + 3^k \subseteq T$. Here

$$2T_h + 3^k = \{2t + 3^k \mid t \in T_h\}.$$

Next, we are going to use mathematical induction to prove: for any $n \in \mathbf{N}^*$, it is always true that $n \in T$.

From the definition of T, we may know that $1 \in T$ is true. Suppose for any $m \in \mathbf{N}^*$, $m < n$, it is always true that $m \in T$. Then we consider the case n.

Case 1. If $2 \mid n$, then $\frac{n}{2} \in T$. Therefore, $n \in 2T \subseteq T$.

Case 2. If $3 \mid n$, then $\frac{n}{3} \in T$. Therefore, $n \in 3T \subseteq T$.

Case 3. If $2 \nmid n$, and $3 \nmid n$, then there exists a $k \in \mathbf{N}^*$ such that $3^k < n < 3^{k+1}$. At this time, $0 < \frac{n-3^k}{2} < \frac{3^{k+1}-3^k}{2} = 3^k < n$. So, $\frac{n-3^k}{2} \in T_{k-1}$. Therefore,

$$n = 2\left(\frac{n-3^k}{2}\right) + 3^k \in 2T_{k-1} + 3^k \subseteq T.$$

So the proposition holds.

31. For any $k \in \mathbf{N}^*$, since f is a surjection, we know that the set $f^{-1}(k) = \{x \mid x \in \mathbf{N}^*, f(x) = k\}$ is a non-empty set. Therefore, by Well-ordering principle, we know that there exists an $m_k \in \mathbf{N}^*$, such that $m_k = \min f^{-1}(k)$.

Next we firstly prove that

$$g(m_k) = k. \quad ①$$

We induct with respect to k. When $k = 1$, from $g(m_1) \leqslant f(m_1) = 1$ and also $g(m_1) \in \mathbf{N}^*$, we may know that $g(m_1) = 1$. Namely, ① is true for $k = 1$.

Now we suppose ① is true for all positive integers that are less than k, namely $g(m_1) = 1, \cdots, g(m_{k-1}) = k - 1$. Then we discuss the value of $g(m_k)$.

Firstly, $g(m_k) \leqslant f(m_k) = k$. Secondly, $\{g(m_1), \cdots, g(m_{k-1})\} = \{1, 2, \cdots, k - 1\}$. Since g is an injective mapping, and from the definition of m_i we know that any two from $m_1, \cdots, m_k$ are different, so $g(m_k) \geqslant k$. Therefore, $g(m_k) = k$. So, ① is true for $k \in \mathbf{N}^*$.

By ① and that g is an injection, we may know that g is a one-on-one correspondence from $\mathbf{N}^*$ to $\mathbf{N}^*$. Now for any $n \in \mathbf{N}^*$, let $g(n) = k$. Since g is a one-on-one correspondence and from ① we know that $n = m_k$, therefore $f(n) = f(m_k) = k$. So $f(n) = g(n)$. The proposition is proved.

32. Consider $n \in \mathbf{N}^*$. Suppose a_n is the nth term, under base 2, of a sequence of positive integers whose 1s are only on even-number positions or whose 1s are only on odd-number positions. We prove: this sequence $\{a_n\}$ meets all conditions.

By making use of binary expressions for positive integers we may know that (1) holds. We only need to prove that (2) also holds. Consider all non-negative integers that are less than 2^{2r}, and they are all $2r$-digit numbers under base 2 (if not up to the number of digits, make the vacancies at the front up by zeros). Among them, there are 2^r numbers whose even-number positions are all zeros, and there are 2^r numbers whose odd-number positions are all zeros. There is only 0 that appears in both types of numbers. Hence, there are exactly $2^{r+1} - 1$ numbers that are less than 2^{2r} in the sequence $\{a_n\}$. Therefore, $a_{2^{r+1}-1} = 2^{2r}$.

For $n \in \mathbf{N}^*$, suppose $2^{r+1} - 1 \leqslant n < 2^{r+2} - 1$, $r \in \mathbf{N}$. Then from the definition of $\{a_n\}$ we know that $a_n \geqslant a_{2^{r+1}-1} = 2^{2r} = \frac{1}{16} \times 2^{2(r+2)} > \frac{n^2}{16}$.

So, there exists a sequence $\{a_n\}$ that meets both conditions.

Bibliography

1. Xiong Bin, Liu Shixiong. High School Math Competition Step-by-step. Wuhan University Press. 2003.
2. Su Chun. A Casual Talk on Techniques When Applying Mathematical Induction. University of Science and Technology of China Press. 1992.
3. Zheng Longxin. Induction and Recurrence. Hubei Educational Press. 1984.
4. Chen Jiasheng, Xu Huifang. Recursive Sequences. Shanghai Educational Publishing House. 1988.
5. Xia Xingguo. Overview of Mathematical Induction. Henan Press of Science and Technology. 1993.
6. Pan Chengdong, Pan Chengbiao. Elementary Number Theory. Peking University Press. 2003.
7. B.J. Venkatachala. Functional Equations. Prism Books Put Ltd. 2002.
8. Feng Zhigang, Xiong Bin. An Introduction to Mathematics Olympiad. Shanghai Science and Technology Educational Publishing House. 2001.
9. Feng Zhigang. Methods and Techniques to Prove Problems by Mathematical Induction. East China Normal University Press. 2005.
10. Feng Zhigang. Elementary Number Theory. Shanghai Science and Technology Educational Publishing House. 2009.